MW00845464

Microbial Genomics and Drug Discovery

Microbial Genomics and Drug Discovery

edited by

Thomas J. Dougherty
Pfizer, Inc.
Groton, Connecticut, U.S.A.

Steven J. Projan
Wyeth Research
Pearl River, New York, U.S.A.

MARCEL DEKKER, INC. NEW YORK • BASEL

Library of Congress Cataloging-in-Publication Data
A catalog record for this book is available from the Library of Congress.

ISBN: 0–8247–4041–6

This book is printed on acid-free paper.

Headquarters
Marcel Dekker, Inc., 270 Madison Avenue, New York, NY 10016, U.S.A.
tel: 212–696–9000; fax: 212–685–4540

Distribution and Customer Service
Marcel Dekker, Inc., Cimarron Road, Monticello, New York 12701, U.S.A.
tel: 800–228–1160; fax: 845–796–1772

Eastern Hemisphere Distribution
Marcel Dekker AG, Hutgasse 4, Postfach 812, CH–4001 Basel, Switzerland
tel: 41–61–260–6300; fax: 41–61–260–6333

World Wide Web
http://www.dekker.com

The publisher offers discounts on this book when ordered in bulk quantities. For more information, write to Special Sales/Professional Marketing at the headquarters address above.

Current printing (last digit):

10 9 8 7 6 5 4 3 2 1

PRINTED IN THE UNITED STATES OF AMERICA

Preface

To date no antibacterial compounds identified by target based screening have advanced into clinical testing much less been used clinically to treat bacterial infections. This well-noted futility has led some to suggest that this approach will never prove productive, that the majority of "novel" targets are not "druggable," and that industry research dollars are better spent elsewhere. After all, even the most stubborn of bacterial infections rarely, if ever, require more than six months of treatment, and the average course of therapy is under seven days. It has been reasoned that it makes more commercial sense to treat chronic conditions requiring years of therapy rather than curing diseases with short courses of treatment. The use of this pharmaceutical industry calculus has long been the rationale used for the chronic underresourcing of antibacterial research programs (not to mention anemic public funding for the study of pathogenic bacteria and antibacterial drug resistance) but this actually flies in the face of commercial and public health realities. Not only are infectious disease therapeutics the second largest source of revenue for pharmaceutical companies (behind cardiovascular drugs), with antibacterial drugs taking the lion's share, but there are also no fewer than six branded products garnering over $1 billion annually, despite fierce generic competition. No other area of therapeutic focus can boast this level of commercial, not to mention therapeutic success.

There is less argument on the need for new antibacterial agents and therapeutic strategies to avoid the emergence and dissemination of resistant bacteria. Part and parcel of the resistance problem are dramatically changing demographics consisting of an aging population with a decline in the robust response to infections. This has produced a population that is becoming increasingly immunosuppressed. In developing countries periodic malnutrition, poor sanitation, the spread of HIV infection, and, almost paradoxically, increasing population density, coupled with the ready availability of cheap, generic antibacterials (often of poor quality), are leading to a mushrooming of resistant bacterial strains. In sum, bacterial diseases in general, and antibiotic-resistant bacterial infections specifically, are increasing in both developing and developed countries. Ironically, even improved therapies for other diseases are contributing to an increase in infectious diseases, e.g. immune suppressive therapies in transplant patients, aggressive chemotherapeutic practices in oncology, and most recently the use of TNF-α antagonists in treating rheumatoid arthritis. Along with this increased susceptibility to infection, it has been well documented that a number of pathogens, including *Streptococcus pneumoniae, Staphylococcus aureus,* and *Mycobacterium tuberculosis,* have developed resistance to a range of antimicrobials at an alarming rate. Yet despite the well-documented commercial opportunity and the clear unmet medical need many large pharmaceutical companies are withdrawing from this field. Contributing to the diminishing interest is the perceived inability of target-based screening to produce commercially successful antibacterials. Some in the industry are now convinced that this approach was doomed from the start.

This is reminiscent of prevailing public and scientific opinion prior to the first successful demonstration of powered flight by the Wright brothers. Prior to their success at Kitty Hawk there were many well-documented attempts at powered flight that literally crashed and burned. Why did Orville and Wilbur Wright succeed where so many others had so visibly, repeatedly—and at times tragically—failed? Three reasons are prominent: (1) *expertise*: especially (as bicycle builders) in the use of durable, lightweight materials, and the ability to learn from the failures of their competitors, (2) *resources*: the Wrights had adequate, although not copious, financial backing, and (3) *a sensitive assay*: the Wrights used the winds of Kitty Hawk to produce additional lift and built what was essentially a glider with an engine. A fourth reason is *persistence*: they were undeterred by the failure of others, patiently working in a systematic fashion to solve problems.

Another important point to consider in target-based screening for antibacterial agents is that, in the pursuit of broad-spectrum agents, the goal is not a new molecular entity that is a potent inhibitor of a single molecular target. Rather, we are searching for inhibitors of a family of related but not identical targets in a wide range of pathogenic bacteria. Several groups have been successful at identifying narrow-spectrum inhibitors with excellent in vitro potency and even efficacy in animal models of infection. But such narrow-spectrum agents are not considered commercially or therapeutically viable because the vast majority of therapeutic applications involve empiric therapy (where an antibacterial drug is administered upon presentation of symptoms but prior to the identification of the

The Wright Brothers Memorial in Kitty Hawk, N.C. The reconstructed hanger and camp building is at the left center. In the distance, the granite Memorial Pylon, completed in 1932, sits atop Kill Devil Hill. The walkway to the right has four stone markers, representing the distances of the four flights made on Dec. 17, 1903. The fourth marker, representing the longest flight of 852 feet, is at the far right of the photograph. The inscription on the Memorial Pylon reads:

IN COMMEMORATION OF THE CONQUEST OF THE AIR
BY THE BROTHERS WILBUR AND ORVILLE WRIGHT
CONCEIVED BY GENIUS, ACHIEVED BY DAUNTLESS
RESOLUTION AND UNCONQUERABLE FAITH

infectious agent). This will be the focus of most antibacterial discovery programs unless and until rapid (read "bedside") diagnostics become widely available. Even then, the enormous expense to bring a drug to market (currently estimated at \$700–800 million) makes a narrow-spectrum agent a difficult commercial proposition. Therefore the assumption made here is that goal of antibacterial drug discovery is a novel class of broad-spectrum agent so the themes of the target classes below are, for the most part, those that are generally well conserved (genetically and structurally) among pathogenic bacteria.

The notion that the mere identification of new targets would quickly result in a flood of novel antibiotics was naive. The rapid delivery of the genomic sequences of pathogenic bacteria over the past eight years raised the expectation that a large number of new targets for therapy would be identified. Indeed, a number of groups have done precisely that using a variety of clever strategies. Genomic target identification certainly has resulted in a more rational and accelerated process in antibiotic target screening. It also suggested several new ways to return to cell-based antimicrobial screens in a targeted approach with increased sensitivity to inhibitors. However, the bottleneck has shifted from target identification to the process of identifying inhibitors that have the potential to become leads, that is, to be modified to drug-like characteristics. This is a difficult and empiric process, involving the balance of continued or improved target inhibition with instilling the proper pharmacological properties and lack of human toxicity in the candidate molecule. The genomics-approach goal is to deliver an increasing number of antibacterial inhibitors against a broader range of targets in the expectation that a small subset of these molecules will indeed be amenable to the necessary pharmacological modifications for safe and effective use as antibiotics.

In this book, a broad range of genomics-based approaches to the problem of identifying novel antibiotic classes are presented. The chapters focus on different aspects of the problems presented in identifying, selecting, and prosecuting novel antibacterial targets. Much like the approach to powered flight a hundred years ago, it is still too early to say which techniques will ultimately yield success. But given the stakes, we must maintain the same faith in our goals as those early pioneers of the air.

Finally, we would like to express our gratitude to the expert authors who agreed to contribute to this endeavor. Their creativity and diligence in the field are clear from the caliber of the work presented. We also thank Anita Lekhwani and Dana Bigelow of Marcel Dekker, Inc., for their efforts in bringing this project forward.

Thomas J. Dougherty
Steven J. Projan

Contents

Contributors

Zhiqiang An Merck Research Laboratories, Merck & Co., Rahway, New Jersey, U.S.A.

John F. Barrett Merck Research Laboratories, Merck & Co., Rahway, New Jersey, U.S.A.

Todd A. Black Chemotherapy and Molecular Genetics, Schering-Plough Research Institute, Kenilworth, New Jersey, U.S.A.

Barry R. Bochner Research & Development, Biolog, Inc., Hayward, California, U.S.A.

Patrick J. Brennan Department of Microbiology, Colorado State University, Fort Collins, Colorado, U.S.A.

Joan K. Brieland Chemotherapy and Molecular Genetics, Schering-Plough Research Institute, Kenilworth, New Jersey, U.S.A.

Dale J. Christensen Karo Bio USA, Inc., Durham, North Carolina, U.S.A.

Guillaume Cottarel Genome Therapeutics Corp., Waltham, Massachusetts, U.S.A.

Dean C. Crick Department of Microbiology, Colorado State University, Fort Collins, Colorado, U.S.A.

Julian Davies Department of Microbiology and Immunology, University of British Columbia, and Cubist Pharmaceuticals, Inc., Vancouver, British Columbia, Canada

Daniel B. Davison Department of Bioinformatics, Bristol-Myers Squibb Pharmaceutical Research Institute Pennington, New Jersey, U.S.A.

Brian A. Dougherty Department of Applied Genomics, Bristol-Myers Squibb Pharmaceutical Research Institute, Wallingford, Connecticut, U.S.A.

Thomas J. Dougherty Molecular Science, Pfizer Global Research and Development, Groton, Connecticut, U.S.A.

Paul M. Dunman Infectious Disease-Antibacterial Research, Wyeth Research, Pearl River, New York, U.S.A.

Claire M. Fraser The Institute for Genome Research (TIGR), Rockville, Maryland, U.S.A.

Steven R. Gill The Institute for Genomic Research (TIGR), Rockville, Maryland, U.S.A.

Christopher P. Gray Morphochem AG, Basel, Switzerland

Jonathan Greene Bioinformatics, Schering-Plough Research Institute, Kenilworth, New Jersey, U.S.A.

Paul T. Hamilton Karo Bio USA, Inc., Durham, North Carolina, U.S.A.

Catherine Hardalo Clinical Research, Infectious Diseases, Schering-Plough Research Institute, Kenilworth, New Jersey, U.S.A.

Peter D. Karp Bioinformatics Research Group, SRI International, Menlo Park, California, U.S.A.

Wolfgang Keck Morphochem AG, Basel, Switzerland

Andrea Marra Antibacterials Discovery, Pfizer Global Research and Development, Groton, Connecticut, U.S.A.

C. Patrick McAtee Lexicon Genetics, Inc., The Woodlands, Texas, U.S.A.

Michael R. McNeil Department of Microbiology, Colorado State University, Fort Collins, Colorado, U.S.A.

Karen E. Nelson The Institute for Genomic Research (TIGR), Rockville, Maryland, U.S.A.

Steven J. Projan Wyeth Research, Pearl River, New York, U.S.A.

Michael J. Pucci Achillion Pharmaceuticals, New Haven, Connecticut, U.S.A.

Richard A. Slayden Department of Microbiology, Colorado State University, Fort Collins, Colorado, U.S.A.

William R. Strohl Merck Research Laboratories, Merck & Co., Rahway, New Jersey, U.S.A.

Scott S. Walker Chemotherapy and Molecular Genetics, Schering-Plough Research Institute, Kenilworth, New Jersey, U.S.A.

Microbial Genomics and Drug Discovery

1

Global Aspects of Antibiotic Resistance

Julian Davies
University of British Columbia, and Cubist Pharmaceuticals, Inc., Vancouver, British Columbia, Canada

Antibiotic resistance is truly a global phenomenon as exemplified by the critical nature of the problem worldwide and the proposed global solutions [1,2]. This is also the case from the perspective of microbial genetics, since the demonstration of the evolution and promiscuous dissemination of resistance genes has provided convincing evidence for the concept of the "global microbial genome." There are many aspects of genetic ecology that contribute to the complexity and enormity of the problem threatening the treatment of infectious disease. The study of resistance and its development is confounded by the fact that, as with most aspects of the study of microbiology, the analysis of antibiotic resistance in bacteria is retrospective; everything has already happened and it is the job of scientists to try to unravel the genetic and biochemical mechanisms that are involved in establishing the phenomenon. Most of the early attempts (as are many of the current studies) to analyze the potential threat of antibiotic resistance development were flawed because of lack of basic knowledge of microbial genetic ecology rather than of a failure to appreciate the problem. Such luminaries of infectious disease and antibiotic research as Alexander Fleming [3] and Maxwell Finland [4] predicted that the lifetime of antibiotic use was likely to be limited (based on their experimental observations) and very much dependent on the patterns of use. What was little known at the time was the ability of microbes to evolve so rapidly and so globally to overcome the threat of extinction.

First, a little history. Antiseptics and disinfectants have been used since the demonstration of the germ theory of disease in the 19th century. Many compounds such as phenolics and metal salts were and still are used in efforts to reduce the risk of infection. In recent years, the promotion of disinfectants and antiseptics (such as triclosan) has fueled a global crusade to eliminate nasty bacteria from any potential human contact [5]. Considering the fact that bacteria have inhabited this planet for approximately 4 billion years and survived many environmental catastrophes it is obvious that this objective will fail; in addition, the extensive use of these agents contributes to the increasing microbial tolerance to antiseptics and concomitantly to other antimicrobials by multiple resistance mechanisms that are now well established and have been always present, but usually ignored.

The first antimicrobials used therapeutically on a large scale were the sulfonamides, which are synthetic compounds discovered in the mid-1930s and still of significant clinical value. Their early use was principally for streptococcal infections and they were remarkably successful, particularly in the early part of the Second World War when used to reduce the incidence of battle wound infections. However, the development of resistance with increasing use was readily apparent even at early stages in their use [6]. The discovery of the first antibiotics (naturally occurring products of soil microbes) led to sweeping changes in the treatment of infectious diseases and had a major impact on medical practice in general, since the ability to cure and prevent bacterial infections became readily available to the medical community. Resistance was increasingly identified but its biological basis poorly understood. In fact, laboratory studies of the development of resistance to penicillin and streptomycin were more often than not directed to studies of bacterial genetics and the nature of mutation; most of this work employed antibiotic concentrations that did not reflect those used in clinical practice. For example, laboratory analyses of streptomycin resistance in *Escherichia coli* (not recognized as a major pathogen at the time) focused on the isolation and properties of strains resistant to concentrations of 100 μg/ml or greater which were obtained following exposure to very high concentrations of drug; for a general survey see [7]. The frequency of appearance of resistant mutants was relatively low (less than 1 in 10^8 cells), which gave a false picture of the clinical situation, and the general impression was that resistance development in treatment would be rare. Nonetheless, in TB sanatoria, *Mycobacterium tuberculosis* strains resistant to clinically significant concentrations of streptomycin often developed during the course of therapy [8]. The writing was on the wall but it seems that few in the medical community really appreciated the problem and spoke to it at that time. The rest is history and Table 1 indicates the sorry milestones that chart the development of antibiotic resistance on a global scale. In retrospect, the failure to recognize and quantitate the global aspects of the problem during the early days of antibiotic use seems absurd; especially since it seems that almost everything happened in the first 2 decades of antibiotic use. Resistance plasmids were identified in 1959 in Japan and in Europe in 1962 [9], but their existence was barely acknowledged in the United States until they were finally identified to be of clinical significance in Boston hospitals in 1966 [10]. Failure to appreciate the

TABLE 1 Some "Landmarks" in the Short History of Antibiotic Use and Antibiotic Resistance

Year	Event
1929	The activity of penicillin identified
1935	Sulfonamides introduced in Europe
1939	Sulfonamide used in the United States
1941	Bacterial penicillinase identified
1944	Streptomycin discovered
1945	Penicillin introduced for general use
1946	Sulfonamide-/streptomycin-resistant strains identified in clinic
1946	Fleming proposes measures to avoid resistance development
1947–1955	Discovery of chloramphenicol, erythromycin, and tetracycline
1955	Large-scale use of chlortetracycline in animal feeds
1958	Vancomycin isolated
1959	6-Aminopenicillanic acid produced
1959	Transferable antibiotic resistance (R-plasmids) characterized in Japan
1960	Methicillin introduced
1961	Methicillin resistance in *Staphylococcus aureus* reported
1962–1966	R-plasmids identified in Europe and the United States
1963	Gentamicin discovered
1965	TEM β-lactamase isolated
1972	Cephamycins discovered
1972	Transformation of *E. coli* with R-plasmid
1973	R-plasmids/resistance genes used to develop recombinant DNA procedures
1974–1975	Transposable elements carrying resistance genes found in bacteria
1988	Plasmid-mediated vancomycin resistance in Enterococci
1989	Neomycin phosphotransferase first approved gene for human gene transfer study
1989	Discovery of integrons and resistance gene cassettes

significance of transferable antibiotic resistance may well have been due to the fact that well-known bacterial geneticists considered it unlikely that sexual mechanisms would contribute to the development of antibiotic resistance in bacteria.

In the year 2000, the development of antibiotic resistance has been recognized as a problem of global proportions and most national and international health authorities are launching initiatives to control the spread of resistance in attempts to restore and maintain the value of antibiotics (Table 2) [1,2]. What other reliable means of treating infectious diseases are available at the present

TABLE 2 Approaches to the Management of Antibiotic Resistance

WHO Global Strategy for the Containment of Antimicrobial Resistance[a]	Antimicrobial Resistance Action Plan Support Act of 2001[b] (Categories of Activity and Top Priority Action Items[c])
A. Patients and the general community	A. Surveillance
B. Prescribers and dispensers	B. Develop and implement procedures for monitoring patterns of antimicrobial drug use in human medicine, agriculture, veterinary medicine, and consumer products
C. Hospitals	C. Prevention and control
D. Use of antimicrobials in food-producing animals	D. Research
E. National Governments and Health Systems	E. Product development
G. International aspects of containing antibiotic resistance	

[a] More than 60 subheadings cover different actions for these topics.
[b] (Thirteen of more than 80 action items are considered top priority).
[c] Introduced by Rep. Sherrod Brown in the 1st Session of the 107th Congress to provide funding for research in the above areas by D.H.H.S.

time? Studies of the epidemiology, biochemical mechanisms, genetic determinants origins, and spread of antibiotic resistance are being actively promoted worldwide. These are all intriguing and challenging problems that involve the discovery and applications of novel genetic and biochemical approaches which have given very good definitions of resistant strains, but very little information on how antibiotic resistance actually develops in clinical practice. Sixty years of extensive antibiotic use and selective pressure cannot be accurately assessed in laboratory isolation. The past 30 years has been a period of remarkable advances in the life sciences that have provided all manner of techniques for the analysis and manipulation of life forms.

Interestingly, studies of antibiotic resistance have provided many of the tools of gene transfer and knowledge of gene evolution that exemplify the almost unlimited variation of genetic behavior in bacteria. It has been known for some time that the processes of antibiotic resistance and bacterial pathogenicity are often linked genetically [11] although the latter is (from an evolutionary point of view) a more ancient process. In the past 60 years, the enhanced genetic traffic driven by selection for antibiotic resistance must also have had an impact on the evolution of mechanisms of pathogenesis and their interspecies spread. Antibiotic resistance determinants are essentially virulence factors; the interspecies movement of resistance genes and their evolution and adaptation by mutagenesis induced by various forms of stress are coincidental with the horizontal transfer and evolution of other virulence factors. One can speculate that bacterial community behavior such as biofilm formation, which plays an important role in the pathogenesis of many infections, has evolved to some extent as a response of microbes to avoid antibiotics under the selection of antibiotic use. The recent work of Ghigo supports this conclusion by showing that sex factors actually promote biofilm formation [12]. Biofilm formation is a good example of the genetic determination of a combined pathogenesis/antibiotic resistance characteristic and it is not the only one. There are now many examples of the genetic linkage between pathogenicity and resistance-determining genes [13].

Pathogenicity islands are complex structures and their transit may be determined by a wide variety of selective pressures. It has been shown that many of these collections of genes of adaptive importance are associated with moveable genetic elements such as transposons, phages, and plasmids; much of this packaging has occurred since 1950 during the period of increasing selection and lateral gene transfer provided by excessive antibiotic usage. It is interesting to note that most of the genetic elements involved are rarely carriers of a single determinant in the form of an antibiotic resistance or pathogenicity gene—multifactorial elements are the rule. There is every reason to believe that the whole may be more than the sum of its parts in these complexes. The appearance of a new antibiotic resistance gene must occur as the result of combined genetic processes and is not likely to be an independent event (e.g., one mutation).

Much has been learned in recent few years concerning the various roles of mutation—the different mechanisms and driving forces—in the development of antibiotic resistance [14]. It was mentioned earlier that the introduction of antibiotic therapy was instituted with complete ignorance of the potential for antibiotic resistance development and the mechanisms involved. Now it is well-established that antibiotic resistance not only occurs but is inevitable! Knowledge of the genetic capabilities and flexibility of microorganisms should be important considerations in planning the therapeutic use of antibiotics. Antibiotic-induced stress contributes to the establishment of hypermutability in bacteria; recent studies of the "mutator" DNA polymerases reveals that this phenomenon may be true for all living organisms [15]. The induction of mutator (or error-prone) polymerases leads to hypermutable states in cells and the presence of antibiotics leads to stress that favors this conditional state. A number of antibiotics such as streptomycin, fluoroquinolones, and nitrofurans have been shown to be also mutagens in their own right. Events associated with gene transfer (conjugation, etc.) are likely to invoke stress responses in the recipient organism and subsequent induced hypermutation contributes to the evolution of both antibiotic resistance and pathogenicity. Shoemaker *et al.* [16] have demonstrated that the gastrointestinal (GI) tracts of mammals are essentially bacterial "bordellos" where genetic exchange is occurring at a very high frequency between many different bacterial species and genera. If mutagenesis occurs concomitantly with transfer of mutator genes, is it any wonder that antibiotic resistance genes so easily evolve to accept new antibiotic substrates? Many of the known resistance gene families were established through cascades of single mutations and their phylogenies characterized; this is especially true in the case of the β-lactamases [17]. It can be assumed that most, if not all, of the mutants were the result of stress-induced hypermutation. In terms of microbial populations, in the "global" GI tract the possibilities for generation, adaptation, and storage of resistance genes predispose the situation for the development of widespread antibiotic resistance. If one ponders bigger environmental issues, what have been the effects on terrestrial and marine microbial ecology as a result of massive antibiotic release? Environmental studies identify the wide distribution of antibiotic-resistant bacteria and associated resistance genes [18] but how extensively have natural microbial ecosystems been changed during a half-century of antibiotic use?

There are many aspects of antibiotic resistance that need further study; for example, the interplay between acquisition and mutation under variable selection is poorly understood at the moment. Some specific examples appear obvious, such as the evolution of the plasmid-borne TEM β-lactamases in response to the introduction of several generations of β-lactam antibiotics [19]. But what might be the effects of these changes on the host? Must the host compensate genetically for each successive change in the evolution of a resistance gene? What occurs on transfer to a heterologous host? Are compensatory mutations (restoration to fitness) associated with every step in the development of a resistant bacterium [20]? How can these events be assessed accurately when resistance plasmids and their associated resistance genes are almost always examined in laboratory-trained bacterial strains such as *E. coli* K12 derivatives, which typically possess many known (and unknown) mutations that attenuate them for use as docile pets in the laboratory? We need to get away from the K12-centric approach; laboratory strains are not good models for the study of the development of resistance. Much of the work on mutation, resistance, and pathogenicity is currently being studied outside of natural genetic contexts. The studies of Cebula *et al.* that led to the identification of hypermutability in bacterial pathogens provide one of the few examples of a "natural context" study [21]. Will the advent of rapid genome sequencing (a bacterial genome in a few hours?) lead to more complete analyses of the evolution of antibiotic resistance pathogens by permitting accurate comparisons of the genetic makeup of sensitive and resistant isolates? What is the wild type? Perhaps the use of isolates from old culture collections [such as the "Murray" strains [22]] will become the neotypes for true analyses of "what makes a pathogen?" or "how does antibiotic resistance develop?" Such studies might give a much better idea of the genetic and physiological parameters involved and provide more accurate and useful answers to these questions. The introduction of novel antibiotic classes into human clinical use in the future should be seized upon as models to study the development of antibiotic resistance in hospital settings as *it happens*; adequate funding should be provided in advance to analyze all components of the response of the microbial community. The information obtained will be invaluable in planning ways to control resistance development. This will be an opportunity to be prospective and not retrospective.

The global development of antibiotic resistance is now a fact of life; it may be controlled but not reversed. Better management of all antibiotic use procedures is much needed. It is clear that the limits (still undefined) of the "global" bacterial genome are critical to an understanding of the microbial world. Any gene of any bacterial species that can be of selective advantage to any other (even distantly related) bacterial species will be subjected to horizontal gene transfer. The global genome implies universal genetic currency independent of biochemical mechanisms. However, the process is complex and it cannot be assumed that an advantageous gene moves directly from bacterium A to bacterium B in a single step; more circuitous routes are probably the norm, especially since there must exist limits to gene transfer in nature. When an antibiotic-resistant bacterium is isolated from a patient in a hospital, it is the only most recent in a chain of events that have taken place within a bacterial community. The concept of the global microbial genome presupposes genetic fluidity and for this reason more emphasis on the study of resistance and virulence genes and their environment in natural isolates is warranted. The principal reservoirs are undoubtedly commensals and since a large percentage of these species are uncultivable, they must be examined for the passage of resistance and virulence determinants by methods other than growth in the laboratory. New techniques will be required; the difficulty will be in linking genotype to species.

Given the above testimonial to the superior genetic and physiological capabilities of microbes one might be led to assume that the problem of antibiotic resistance has no solution. If antibiotics continue to be used in the manner employed currently, there is probably no solution. But there are realistic approaches that will lead to reduction and delay of the appearance of resistant strains in the microbial population. Comprehensive programs of action have been recommended by many groups [23,24]; these measures must be actively pursued locally and globally. As scientists, we remain at a disadvantage in that a complete genetic and physiological picture of the development of antibiotic resistance is still not available. It is very clear that there are many more questions than answers—is it not time to address this imbalance? More studies are needed, both retrospective and prospective, to learn how bacteria respond and survive the stress of antibiotic exposure and how they adapt to changing antibiotic therapies in human hosts.

One place to start might be to look into the functions of so-called antibiotics in natural microbial communities. A better knowledge of the small molecule biology of microbial communities will likely provide a better notion of the range of the responses of bacteria and bacterial populations to antibiotics and even lead to novel approaches to the discovery of useful growth inhibitors. The best antibiotics are biologically active natural products and they should be examined naturally.

REFERENCES

1. Available at www.who.int/emc/amr.html.
2. U.S. Congress. *Public Health Action Plan to Combat Antimicrobial Resistance*. Washington DC: Department of Health and Human Services.
3. A Fleming. In: PA Hunter, GK Darby, NJ Russell, eds. *Fifty Years of Antimicrobials: Past Perspectives and Future Trends*. Cambridge, UK: Cambridge University Press, pp 1–18, 1946.
4. M Finland. *Ann. NY. Acad. Sci. 182*:5–20, 1971.
5. SB Levy. *Emerg. Infect. Dis. 7*:512–515, 2001.
6. DS Damrosch. *J. Am. Med. Assoc. 130*:124–128, 1946.
7. W Hayes. *The Genetics of Bacteria and their Viruses. 2nd ed.* Edinburgh: Blackwell Scientific, 1968.
8. GP Youmans, AG Karlson. *Am. Rev. Tuberc. 55*:529–544, 1947.
9. J Davies. *Genetics 139*:1465–1468, 1995.
10. DH Smith. *N. Engl. J. Med. 275*:626–630, 1966.
11. KN Timmis, I Gonzales-Carrero, T Sekizaki, F Rojo. *J. Antimicrob Chemother 18 (suppl C)*: 1–12, 1986.
12. J-M Ghigo. *Nature 412*:442–445, 2001.
13. C Chu, C-H Chiu, W-Y Wu, C-H Chu, T-P Liu, JT Ou. *Antimicrob Agents Chemother 45*:2299–2303, 2001.
14. GJ McKenzie, SM Rosenberg. *Curr. Opin. Microbiol. 4*:586–594, 2001.
15. F Taddei, I Matic, B Godelle, M Radman. *Trends. Microbiol. 5*:427–428, 1997.
16. NB Shoemaker, H Vlamakis, K Hayes, AA Salyers. *Appl. Environ. Microbiol. 67*: 561–568, 2001.
17. K Bush, GA Jacoby, AA Medeiros. *Antimicrob Agents Chemother 39*:1211–1233, 1995.
18. VC Nwosu. *Res. Microbiol. 52*:421–430, 2001.
19. T Palzkill. *ASM. News 64*:90–95, 1998.
20. DI Andersson, BR Levin. *Curr. Opin. Microbiol. 2*:489–493, 1999.
21. JE LeClerc, B Li, WL Payne, TA Cebula. *Science 274*:1208–1211, 1996.
22. VM Hughes, N Datta. *Nature 302*:725–726, 1983.
23. RJ Williams, DL Heymann. *Science 279*:1153–1154, 1998.
24. ML Cohen. *Nature 406*:762–767, 2000.

2

Genomics of Bacterial Pathogens

Steven R. Gill, Karen E. Nelson, and
Claire M. Fraser
The Institute for Genomic Research, Rockville, Maryland, U.S.A.

INTRODUCTION

The completion of the whole-genome sequencing of *Haemophilus influenzae* [1] in 1995 launched the era of microbial genomics. As well as being the first free-living organism to have all its genetic information deciphered, *H. influenzae* is a major pathogen that would give tremendous insights into how organisms cause disease. Since the release of *H. influenzae*, an additional 22 microbial pathogen genomes have been completely sequenced, and another 52 pathogen genome-sequencing projects are known to be in progress (see Table 1). The diversity of pathogens that have been chosen for whole-genome sequencing has and continues to allow for the identification of new antimicrobial targets, vaccine candidates, and an increased understanding of how these organisms cause disease. Complete genome sequences also allow for comparative studies of pathogenicity, often revealing major factors associated with virulence such as pathways for capsule synthesis and novel secreted proteins (e.g. for *Bacillus anthracis*, *Pseudomonas aeruginosa*, *Streptococcus pneumoniae*, and *Neisseria meningitidis*). Revelations into genome composition, genome rearrangements, and organizational differences across these species are also being gleaned.

CONDUCTING A GENOME PROJECT

Sequencing

Although a variety of methods were initially used for generating complete microbial genome sequences, random shotgun sequencing has become the method of

TABLE 1 Completed and Ongoing Genome Projects for Pathogens

Organism	Size (Mb)	Relevance/Disease Caused
COMPLETED		
Bacillus anthracis	4.5	Anthrax
Borrelia burgdorferi	1.4	Lyme disease
Brucella suis	3.3	Brucellosis
Brucella mellitensis	3.3	Malta Fever
Campylobacter jejuni	1.6	Bacterial food-borne diarrheal *disease*
Chlamydia muridarum	1.1	Respiratory infections in mice
Chlamydia pneumoniae	1.0	Pnemonia, arthersclerosis
Chlamydia trachomatis	1.0	STD, nongonococcal urethritis, eye infections
Escherichia coli O157:H7	5.5	Diarrhea, hemorrhagic colitis
Haemophilus influenzae	1.8	Meningitis
Helicobacter pylori	1.7	Gastric ulcers
Mycobacterium leprae	3.3	Leprosy
Mycobacterium tubercolosis	4.4	Tubercolosis
Mycoplasma genitalium	0.6	Nongonococcol urethritis
Neisseria meningitidis	2.3	Meningitis, pharyngitis, septicaemia
Staphylococcus aureus	2.8	Toxic Shock Syndrome
Streptococcus agalactiae	2.2	Pneumonia, meningitis
Streptococcus mutans	2.0	Dental caries
Streptococcus pneumoniae	2.2	Meningitis, pneumonia, septicaemia
Streptococcus pyogenes	1.8	Rheumatic fever
Treponema pallidum	1.1	Venereal syphilis
Ureaplasma urealyticum	0.8	Mucosal disease
Vibrio cholerae	2.5	Cholerae
IN PROGRESS		
Actinobacillus actinomy	2.2	Periodontal pathogen
Bacteroides forsythus	2.2	Periodontal pathogen
Bacteroides fragilis	5.3	Opportunistic pathogen
Bartonella henselae	2.0	Cat scratch disease
Bordetella parpertussis	3.9	Whooping cough
Bordetella pertussis	3.9	Whooping cough
Burkholderia cepacia	8.0	Opportunistic pathogen in CF *patients*
Fusobacterium nucleatum	2.4	Periodontal pathogen
Porphorymonas gingivalis	2.2	Oral infections
Campylobacter coli	1.6	Acute bacterial diarrhea
Campylobacter upsaliensis	1.6	Human enteropathogen, *spontaneous abortions*
Chlamydophila felis	1.2	Feline conjunctivitis

(*continued*)

TABLE 1 Continued

Organism	Size (Mb)	Relevance/Disease Caused
Clostridium tetani	4.4	Tetanus
Clostridium perfringens	3.0	Food poisoning, gas gangrene
Corynebacterium diphtheriae	2.5	Diphtheria
Coxiella burnetii	2.1	q Fever
Dichelobacter nodosusi	1.6	Footrot in sheep
Ehrlichia chaffeensis	1.2	Human monocytotropic ehrlichiosis
Enterococcus faecalis	3.0	Bacteremia, endocarditis
Enterococcus faecium	2.8	Bacteremia, endocarditis
Francisella tularensis	2.0	Tularaemia
Fusobacterium nucleatum	2.4	Dental pathogen
Haemophilus ducreyi	1.8	Genital ulcers
Klebsiella pneumoniae	6.0	Pneumonia
Legionella pneumophila	4.0	Legionnaires' disease
Leptospira interrogans	4.8	Leptospirosis
Listeria monocytogenes	2.8	Food-borne epidemics, listeriosis
Mannheimia haemolytica A	12.7	Bovine pneumonic pasteurellosis
Mycobacterium avium	4.7	Crohn's disease
Mycobacterium bovis	4.4	Bovine tuberculosis
Mycoplasma hyopneumoniae	0.9	Porcine pneumonia
Mycoplasma mycoides	1.3	Bovine pleuropneumonia
Mycoplasma pulmonis	1.0	Murine respiratory diseases
Neisseria gonorrhoeae	2.2	Gonorrhoea
Pseudomonas pseudomallei	6.0	Meliodosis
Porphyromonas gingivalis	2.2	Periodontal pathogen
Prevotella intermedia	3.8	Periodontal pathogen
Rickettsia conorii	1.2	Mediterranean spotted fever
Rickettsia prowazekii	1.1	Typhus
Salmonella enterica	4.5	Gastrointestinal infection
Salmonella enteritidis	4.6	Gastrointestinal infection
Salmonella paratyphi	4.6	Paratyphoid fever
Salmonella typhi	5.0	Typhoid fever
Salmonella typhimurium	4.5	Gastrointestinal infection
Staphylococcus epidermidis	2.6	Nosocomial pathogen
Shigella flexneri 2a	4.7	Acute bacillary dysentery
Streptococcus gordonii	2.2	Oral colonizer
Streptococcus mitis	2.2	Oral colonizer
Streptococcus sobrinus	2.2	Oral colonizer
Treponema denticola	3.0	Oral infections
Yersinia pestis	4.7	Bubonic plague
Yersinia pseudotuberculosis	4.3	Acute ileitis

choice. This now-standardized procedure has been used on organisms that range in genome size and G + C content and for organisms with multiple extrachromosomal elements. Total DNA of the organism of choice is randomly sheared and cloned into a plasmid, and the ends of the clones are sequenced to a predetermined level of coverage that represents the entire genome. Random libraries, with few no-insert and chimeric clones, are critical for a good representation of the entire genome during the random sequencing phase.

Following the random sequencing phase, the sequences are assembled into contigs, and any remaining unsequenced regions of the DNA are closed by a combination of methods that include walking spanning clones, sequencing a polymerase chain reaction (PCR) product generated from primers designed at the ends of the contigs, or combinatorial/multiplex PCR. Direct walking on bacterial DNA can also be used to close these gaps. All repetitive sequence regions, including IS elements, ribosomal RNA regions, or transposons, are confirmed by walking spanning clones across these regions.

Data Mining of Pathogen Genomes

Once the genome sequence is closed, bioinformatic analyses of the genome sequence become essential for interpreting the basic biology of the species. Bioinformatic analysis includes identification of all open reading frames (ORFs) and other elements (tRNA, rRNA, repeated sequences, etc.) in the genome. These analyses often extend to the identification of intergenic regions and novel features on the genome, including nucleotide biases, origins of replication, putative regions of horizontal gene transfer, repeat structures, insertion elements, and plasmids. Effective automated gene identification can be accomplished with programs that employ Hidden Markov models (HMMs) or Interpolated Markov models [e.g., GLIMMER [2]] with biological name assignments and functions being made by a combination of computer programs and human curation. BLAST [3] or FASTA [4] searches against sequence databases, and comparisons with homologous families of proteins, including HMMs, Pfams, and clusters of orthologous groups (COGs), aid in the process of making functional predictions.

A closer examination of the genome sequence can allow for a reconstruction and description of the basic biology of the organism. For example, we have been able to reconstruct biochemical profiles for a number of microorganisms, most recently for the pathogens *Streptococcus pneumoniae* type 4 [5] and *Staphylococcus aureus* and *Staphylococcus epidermidis* (Gill *et al.*, in preparation). *Streptococcus pneumoniae* is the causative agent of life-threatening diseases, including pneumonia, bacteremia, and meningitis, with high morbidity and mortality rates among young children and elderly people. It is also the leading cause of otitis media and sinusitis [6]. Capsular types are considered to be major virulent determinants across *Pneumococcal* strains (and other bacterial species, including *B.*

anthracis, *H. influenzae*, and *P. aeruginosa*) and are also known to differ significantly. Detailed analysis of the genome sequence allowed for the reconstruction of the putative pathway for biosynthesis of the type 4 capsule, as well as for the identification of all surface exposed proteins and hence possible external virulence factors.

The Institute for Genomic Research (TIGR) has developed a database, the Comprehensive Microbial Resource (CMR) (available at www.tigr.org/), that enables inter- and intragenomic comparisons of microbial genomes. The CMR contains tools for querying a database that incorporates a detailed curated genome dataset from each of TIGR's microbial genome projects as well as the original annotation and further automated annotation by TIGR for all of the non-TIGR microbial genome sequencing projects. This allows the user to compare genomes based on role categories, protein families, best matches, and so on, and enables complex queries based on a variety of features.

COMPARATIVE GENOMICS WITHIN AND ACROSS PATHOGENIC SPECIES

Sequences from several related pathogens are currently available, and in many cases several strains of particular pathogenic species can be accessed via public databases. Comparative genomic studies have allowed for the identification of major and minor differences between related strains. For example, similar overall genomic organization has been found between *Helicobacter pylori* strains. Approximately 7% of the genes were identified as specific to each strain, with many being clustered in a single hypervariable region [7]. Comparative approaches based on completed genomes have also been used to identify "zones of plasticity" in related *Chlamydia* species that are highly conserved throughout the rest of the genome [8]. In closely related strains of *N. meningitidis*, 239 ORFs in strain MRC58 were absent from strain Z2491. Two hundred eight of the ORFs from Z2491 were absent from MRC58, most of which were characterized as virulence associated genes and could be associated with atypical regions thought to be acquired by gene transfer [9].

APPLICATION OF FUNCTIONAL GENOMICS FOR DETERMINATION OF GENE FUNCTION

Many genome-based drug discovery-screening strategies rely on the biochemical activity of gene products. Therefore, accurate determination of gene function is essential for construction of these screens. A significant amount of effort has been directed at developing *in silico* tools for assignment of gene function. These tools no longer rely solely on BLAST [3] or FASTA [4] linear sequence alignments with known genes, but rather use more sophisticated algorithms for searches, such as HMMs [two examples being the Pfam [10] and TIGRFAM [11] sets], PSI-BLAST [12], and COGs [13]. Additional methods such as threading [14] attempt to take three-dimensional (3D) relationships, such as protein-fold recognition, into account. Despite these new approaches, many genes (or proteins) identified in a genome sequence have no known function and are commonly referred to as hypothetical or conserved hypothetical proteins. In a typical bacterial genome, these genes with no identified function constitute 25–50% of the total gene content. Because many of these genes likely play critical roles in cell function, they are not only of biological interest, but may also serve as targets for genome-based drug discovery.

The era of bacterial genome sequencing has been paralleled by the development of high-throughput genome-based technologies, which have revolutionized our approaches in molecular biology and epidemiology. The expression level of thousands of genes can now be monitored and the identity of thousands of proteins from bacteria can be determined. Application of these technologies to bacterial genome sequences has enabled researchers to (1) identify genome sequence differences between closely related strains, (2) determine function of genes whose function could not be assigned by available *in silico* technology, (3) identify interacting genes and proteins, (4) identify regulatory gene networks, (5) identify genes essential for pathogenicity and survival in the host animal, and (6) develop protein interaction maps. These strategies include subtractive hybridization [15,16], proteomics or functional proteomics [17], two-hybrid protein–protein interaction methods [18], large-scale mutagenesis studies [19], and microarray expression technology [20].

Other genome-based technologies which establish relationships between genes and pathogenicity fall into two broad classes: [1] signature tagged mutagenesis (STM) [21], which uses comparative hybridization to identify genes essential for survival in the host and to isolate mutants unable to survive in specified environmental conditions; and [2] *in vivo* expression technology (IVET) and recombinant *in vivo* expression technology (RIVET) [22,23], which use a promoter-trap strategy to identify genes whose expression is induced in a host infection. The requirement of an animal host or model for both STM and IVET has led to the development of genomic-based approaches which bypass the need for an animal host. One of these approaches, *in vivo* induced antigen technology (IVIAT), uses human sera from human patients to probe for genes specifically expressed *in vivo* [24]. Several of these technologies are discussed in detail below.

Subtractive Hybridization

Subtractive hybridization (SH) is an efficient alternative to whole-genome sequencing of closely related strains that can give insight into strain differences as relates to different pathogenicities. In the SH method, DNA's from the two species in question are fully digested with a frequent cutter to yield fragments of average size between 250 and 2000 bp. Subtracted DNA's can ultimately reveal sequences from clone libraries that are enriched for unique sequences from the two genomes that are being compared. The SH technique has been used to successfully identify unique regions in closely related strains of *H. pylor* [15] and has allowed for the identification of genes that are absent from related species that may contribute to observed differences in pathogenicity. In a similar study, subtractive hybridization was used to compare the related but divergent *Esherichia coli* and *Salmonella typhimurium* genome sequences [16].

DNA Microarrays

In terms of speed and high throughput, high-density oligonucleotide and DNA microarrays and proteomic analysis are becoming the primary tools for global analysis of gene expression and protein complexity in pathogenic bacteria. In a bacterial DNA microarray, PCR fragments or oligonucleotides representing the entire complement of genes contained within a bacterial genome are immobilized on a solid support. These DNA microarrays can be used in two general approaches to identify drug candidates effective in controlling bacterial growth and infectivity. In the first approach, the DNA microarrays are used to monitor bacterial gene expression in response to selected environmental conditions or after exposure to potential drug candidates. This microarray analysis of gene expression at the genomic level enables investigators to understand the mechanistic basis of many drugs and design new drugs based on newly identified regulatory pathways or networks. For example, growth and virulence of many bacterial pathogens is controlled by several regulatory factors, some of which have been recently identified from the genome sequences. Whole-genome analysis of gene expression will enable us to determine what role these regulatory factors play in virulence and develop new drugs which inhibit their activity. This approach has recently been used with *H. influenzae* to explore transcriptional responses triggered by exposure to novobiocin or ciprofloxacin [25] and in *S. pneumoniae* to investigate transcriptional responses of quorum-sensing [26] and competence [27] systems.

Another powerful use of DNA microarrays is as a comparative genomics tool for identifying genetic differences between related bacterial isolates. For example, DNA microarrays of the *Mycobacterium tuberculosis* genome were used to identify differences between *M. tuberculosis* [28] and attenuated bacille Calmette-Guérin (BCG) strains of *Mycobacterium bovis*, which are currently used as a vaccine to prevent tuberculosis. Continual passage of the BCG isolates has led to a dispersion of substrains with differing levels of virulence and immunogenicity. Genomic DNA from several BCG isolates was used for hybridization experiments to screen for ORFs or genes deleted from BCG strains when compared to *M. tuberculosis*. Sixteen regions, varying in length from 1,900 to 12,700 bp, were deleted in BCG. Genes identified with these deleted regions encode likely virulence factors which are prime candidates for generating a more effective BCG vaccine. In a similar fashion, DNA microarrays have been used to investigate genetic diversity in *S. pneumoniae* [29], *H. pylori* [30], and *S. aureus* [31]. Because it is critical that therapeutic targets be conserved among the relevant natural isolates, genes conserved among clinical strains of these pathogens represent potential therapeutic targets.

An alternative to investigating gene expression in the pathogen is gene expression of the host in response to challenges with the pathogen [32]. DNA microarrays containing all ORFs from the genomes of several models of infectious disease (in human, rat, and mouse) are now, or imminently, available for these experiments. In this case, elucidation of the host response to a pathogen could help determine the most potentially effective drug candidate.

Proteomics

The development of DNA microarrays for analysis of gene expression has been paralleled by the development of proteomics, a complementary technology used to analyze global patterns of gene expression at the protein level. The technology developed for proteomic analysis is complex and, as with microarrays, continues to evolve. The most dominant current technology makes use of high-resolution two-dimensional gels followed by sequence analysis of the resolved proteins using matrix-assisted laser desorption ionization (MALDI) time-of-flight (TOF) mass spectrometry [33,34]. When compared to DNA microarrays, proteomic analysis of a cell or bacteria is arguably a more accurate reflection of the physiological state of that cell or bacteria. This is because proteins are frequently the functional molecules in a cell and, therefore, are likely to reflect differences in gene expression. Furthermore, the level of a particular gene transcript does not necessarily reflect the number of functional protein molecules in a cell [35]. In addition to the relative abundance of each protein, proteomic analysis also enables investigators to explore protein turnover and post translational modifications.

Proteomic analysis can be applied to the study of pathogenic bacteria and identification of drug targets in several ways. For example, fractionation of *S. aureus* into separate protein components (e.g., cell-wall proteins) followed by proteomic analysis has enabled investigators to identify cell wall proteins, which are likely vaccine candidates [36]. A similar comparison of culture supernatants from wild-type and mutant group A *Streptococcus* facilitated the identification of previously undescribed immunogenic extracellular proteins [37]. Proteomic analysis can also be used to investigate regulatory and postregulatory effects on pathogenic bacteria grown in the presence of selected antibiotics.

Gene Traps and Knockouts

With the availability of complete bacterial genome sequences, we are now in a position to determine what genes are essential for cellular life and what gene functions are conserved between pathogens. These genes represent likely candidate targets for drug and vaccine development. Identification of these genes can be achieved by large scale gene knockout studies using saturation transposon mutagenesis with sequence identification of insert sites, targeted gene knockouts of sequenced bacterial genomes, IVET, IVIAT, and STM [38,39,40,19]. A good example of this approach is the random transposon mutagenesis of *Mycoplasma genitalium* and *Mycoplasma pneumoniae*, which harbor among the smallest of known bacterial genomes [19]. This approach demonstrated that of the 480 protein-encoding genes in *M. genitalium*, approximately 265 to 330 were essential for growth under laboratory conditions and therefore provide an estimate of the minimal genome required to sustain life. In a similar manner, application of IVET and STM tools to pathogens such as *S. typhimurium*, *Vibrio cholerae*, and *S. aureus* have enabled investigators to begin identification of those genes essential to growth in the complex environmental conditions encountered within the infected host [39].

Genomic Two-Hybrid Interactions

The yeast two-hybrid approach was first developed as a method to detect the interaction between two proteins by transcriptional activation of one or several reporter genes [41]. It has since become the method of choice for identifying pairs of proteins that physically associate with one another and placing these proteins within a biological context toward the goal of understanding their functional roles. Because two-hybrid screens are simple sensitive, and adaptable to high-throughput methods, they are now being adapted to investigating protein–protein interactions on a genomic scale. For example, two-hybrid screens can be designed to explore protein–protein interactions of all proteins encoded by an organism and develop proteome-wide protein interaction maps where totally unknown proteins are partnered with annotated proteins in the same functional category. A direct consequence of identifying physical interactions between proteins is the identification of their interacting domains. These interacting domains represent a first step toward the development of assays for modulation of protein–protein interactions that are applicable to new drug design.

The first genome-scale analysis using the two-hybrid approach was done with the T7 bacteriophage [42]. Subsequent efforts have focused on developing whole-genome interaction maps of *Saccharomyces cerevisiae* [43] and *H. pylori* [18]. Alternative two-hybrid systems have been developed for *E. coli* [44,45]. Because these systems have the advantage of a much shorter generation time in *E. coli*, more flexible molecular biology techniques, and higher transformation efficiencies, they are more adaptable to high-throughput methods required for whole-genome sequencing.

Genotyping and Identification of Single Nucleotide Polymorphisms in Bacterial Genomes

Complete genome sequencing of multiple isolates of closely related bacterial pathogens avails us with the opportunity to explore these sequences for the presence of single nucleotide polymorphisms (SNPs), insertions and deletions, repeats, and tandem repeats, in addition to identifying regions of exact match between the multiple genomes. Identification of SNPs or other differences between multiple genomes can be done using software such as MUMmer [46], which was developed for pairwise alignment and comparison of very large segments of DNA (millions of basepairs in length). This approach has recently been used to identify sequence polymorphisms and SNPs in the sequenced isolates of *M. tuberculosis* (H37Rv laboratory strain and CDC1551 clinical isolate) [47]. A total of 74 large sequence insertions and deletions were identified between the *Mycobacterium* genomes, many of which are found in genes likely contributing to virulence. In a similar manner, MUMmer can be used to identify sequence polymorphisms among the sequenced isolates of other bacterial pathogens such as *S. pneumoniae* and *S. aureus*. Alignment of these multiple genomes would allow us to identify gene families with potentially important pathogenic characteristics. For example, variable or hypervariable regions in genes encoding cell-wall-associated virulence factors may represent a significant source of antigenic variation that plays a role in the immune response of the host. These variable regions may also represent potential targets for novel therapeutic agents.

Targeted SNP analysis and comparison of these hypervariable regions among multiple clinical isolates of bacterial pathogens may also be used to investigate genome diversity and the genetic basis of virulence. The SNP data can be used to construct phylogenetic trees to investigate evolutionary relationships which will perhaps lead to a better understanding of acquisition of virulence factors and antibiotic resistance genes between community and nosocomial isolates.

The technology for the identification of SNPs has been driven by the utility of SNPs as genetic markers in the human genome for establishing linkage and as indicators of genetic diseases. Identification of the greater than 3 million estimated SNPs in the human genome will require the development of accurate, high-throughput, and economical methods. While genotyping technologies are evolving at a rapid pace, several tools are currently being used in academic or industry settings. These include (1) hybridization to high-density oligonucleotide arrays (48); (2) MALDI-TOF mass spectrometry, which differentiates genotypes based on the mass of variant DNA sequence [33,49,50]; and (3) PCR-based approaches [51,52]. Because of its high sensitivity and capacity for high-level multiplexing, MALDI-TOF demonstrates the greatest promise for high-throughput, economical genotyping.

Structural Genomics

The explosion of genome sequence and protein sequence data has led to the growth of the field of structural genomics, which utilizes X-ray crystallography or nuclear magnetic resonance (NMR) spectroscopy for determination of protein structure or fold. The comparison of the sequence of a protein of known function with the three-dimensional structure of that protein allows investigators to begin correlating structure with function. While the overall goal of structural genomics is to provide three-dimensional models for every known protein, current projects are focused on determining the structures of all possible protein folds in nature [53,54]. The number of compact globular protein folds in nature is relatively small. According to current estimates, there are 1000–5000 distinct, stable polypeptide chain folds in nature [55]. Large-scale determination of all protein structures and folds will facilitate the development of new computational tools for assigning proteins folds based on primary sequence and for comparative protein structure modeling. Application of these new computational approaches to complete genome sequences will enable investigators to determine functions of the remaining approximately 20–50% unknowns in each bacterial genome sequence. The approach relies on cloning of candidate genes in suitable expression vectors, followed by overexpression and purification of the protein. While much effort has been focused on the technologies for production and purification of proteins for structural analysis [56], the study of membrane proteins remains problematic because they are typically difficult to purify and are often too large for NMR-based approaches.

GENOMIC APPROACHES TO VACCINE DEVELOPMENT

Technology for vaccine development has historically relied on the use of attenuated or killed microorganisms, bacterial toxins, and polysaccharide—carrier protein conjugates. However, the emergence of new pathogens for which traditional vaccine development approaches have failed, and the desire to use specific cloned antigens that can be tested for safety and efficacy, have led to the utilization of genomics-based approaches for current vaccine development. Bacterial proteins considered to be likely candidates for vaccine development include surface-associated exposed proteins and secreted bacterial proteins or virulence factors. While many of these proteins may be antigenic, only a fraction may be able to stimulate a protective immune response. The application of genomic technology significantly shortens the time required both for identification of antigens and their utility as protective immunogens.

Genomics tools that are being utilized for vaccine development include bioinformatics, proteomics, microarrays, IVET/STM technologies, and emerging expression/display methods. In one strategy, the complete genome sequence of a bacterial pathogen can be mined using *in silico* methods to identify genes encoding predicted surface-associated antigens [57,58]. Alternatively, surface-associated antigens can be selected experimentally by cellular fractionation of the bacteria into cell wall proteins and subsequent characterization of these proteins by two-dimensional gel electrophoresis and mass spectrometry. Genes encoding proteins identified by either approach can then be cloned into expression vectors for subsequent expression and purification of each protein and further determination of its suitability as an antigen.

The *in silico* approach was used by Pizza and co-workers [58] to identify 600 surface-associated proteins from the genome of serogroup B *Meningococcus*. Of these 600 candidate antigens, 350 were successfully expressed in *E. coli*, purified, and used to immunize mice for production of antisera. Antiserum against 25 of these candidates was found to demonstrate bactericidal activity, which correlates with efficacy against meningococcal infections. In addition, several bactericidal candidates contain multiple protective epitopes which are conserved in most MenB isolates, making these optimal vaccine candidates. This approach (Figure 1) is currently being used to identify potential vaccine candidates from

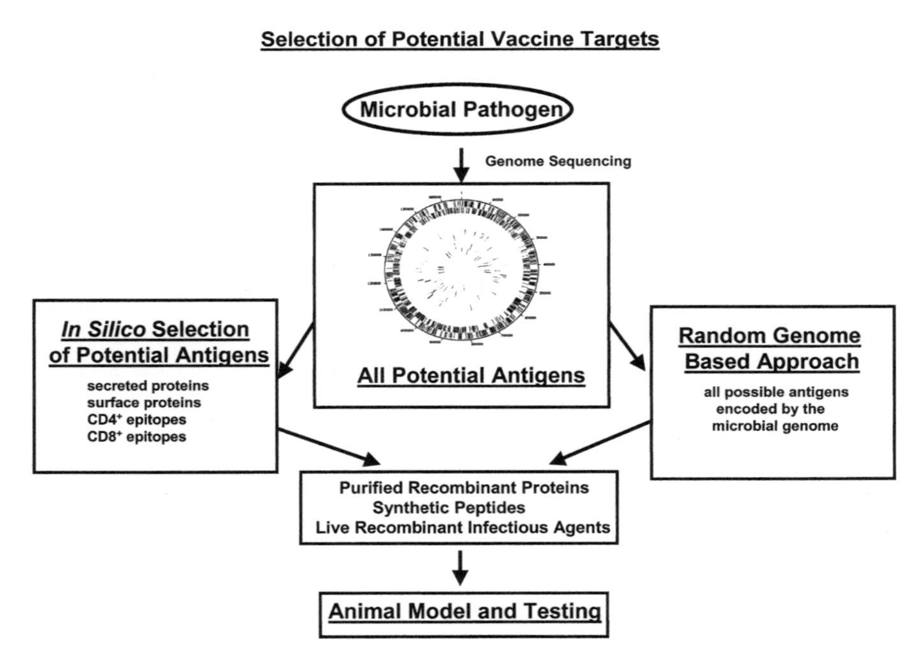

FIGURE 1 Diagram depicting application of complete bacterial genome sequence for vaccine development.

additional bacterial pathogens, including *Chlamydia pneumoniae* and Group B *Streptococcus*.

Application of the *in silico* approach to identify surface-associated or secreted proteins is limited both by the ability of predictive algorithms to identify potential antigenic candidates and by the need for expression and purification of the protein in a heterologous host. An alternative approach developed by Etz and workers [36] utilizes genome-derived peptide libraries to provide a comprehensive antigenic profile of the pathogen. In this method, genomic DNA from the selected bacterial pathogen is digested into small fragments (50–300 bp) that are cloned and expressed in *E. coli* surface display vectors. Human antisera from patients recovering from an infection of the bacterial pathogen being studied is collected and used to immunoselect surface-displayed epitopes which represent candidates that are recognized by the human immune system. This method has the advantage of expressing all possible antigens encoded by the bacterial genome and selecting for those that are likely bactericidal. Using this approach, Etz and workers [36] have identified more than 60 antigenic proteins from the genome of *S. aureus* and are currently working on identification of antigens from *S. epidermidis* and other bacterial pathogens.

While these two genomics based methods are directed toward identification of B-cell epitopes that direct humoral immunity, protection against other microorganisms will require identification of T-cell epitopes capable of stimulating cellular immunity. Efforts to develop vaccines against *Mycobacterium leprae* [59] and *Plasmodium falciparum* [60–62] have focused on the identification of both B-cell and T-cell epitopes which can effectively stimulate both humoral and cellular immunity. For example, Shi and co-workers [62] have synthesized a multivalent, multistage malarial vaccine candidate that contains 12 B-cell and 9 T-cell epitopes derived from 9 life-stage-specific antigens of *P. falciparum*. Identification of additional conserved B- and T-cell epitopes from the *P. falciparum* genome sequence can now be used to assemble additional multivalent antigens capable of eliciting long-lasting immunity.

PERSPECTIVES

The application of genomics and new functional genomics technologies to the study of bacterial pathogens has allowed us to make significant strides in our understanding of their physiology and virulence mechanisms and the design of effective antimicrobials. Complete genome sequencing of additional pathogen genomes and development of new functional genomics tools to exploit the sequence information will only accelerate the pace of discovery. However, our discoveries involving known pathogens may appear miniscule when one considers that greater than 99% of all bacterial species remain to be cultivated, including bacterial inhabitants of plants and animals and environmental bacteria in terrestrial or aquatic ecosystems. By exploring this vast realm of microbial diversity, we stand to gain significant amounts of information on how organisms cause disease and some of the genes—particularly the conserved hypotheticals and uniques—that may be involved in disease formation. Many of these "unculturables" may represent the infectious agents responsible for acute and chronic diseases, such as Crohn's disease and ulcerative colitis, where no infectious agent has been identified [63]. Identification and establishment of physiological requirements for growth of these organisms under laboratory conditions will allow further investigation of these species.

Genomic-based approaches that have been successfully used in the identification of novel genes in unculturable environmental species have included ribosomal-based analyses in combination with sequencing of large segments of DNA cloned in bacterial artificial chromosomes. Bacterial artificial chromosomes can be sequenced and processed through an annotation pipeline method similar to what is currently accomplished with microbial genomes from cultivated species [64,65]. Using these techniques, Beja and workers [64,65] have shown that archaeal-like rhodopsins are broadly distributed and may contribute significantly to light-driven energy generation. New virulence factors, and mechanisms for causing disease, could be identified from unculturable human pathogens in a similar fashion.

The availability of complete genomes from a number of species that have pathogenic and nonpathogenic relatives or members that function as both plant and animal pathogens will also allow for more detailed comparative studies, which may reveal what genes contribute to bacterial pathogenicity. For example, comparative genome analysis of the pathogen *Pseudomonas aeruginosa* with the nonpathogenic *Pseudomonas putida* has revealed that 15% of the *P. aeruginosa* genome is unique and may contain virulence factors contributing to pathogenicity [66]. Similar comparisons can be made between two Staphylococcal species: *S. aureus*, which is an aggressive nosocomial pathogen, and *S. epidermidis*, which is an opportunistic nosocomial pathogen. Genes that are associated in pathogens to the exclusion of their close relatives should be good targets for increasing current understanding of previously undescribed factors that contribute to pathogenicity of these species.

REFERENCES

1. RD Fleischmann, MD Adams, O White, RA Clayton, EF Kirkness, AR Kerlavage, CJ Bult, JF Tomb, BA Dougherty, JM Merrick, K McKenney, G Sutton, W FitzHugh, C Fields, JD Gocayne, J Scott, R Shirley, L-I Liu, A Glodek, JM Kelley, JF Weidman, CA Phillips, T Spriggs, E Hedblom, MD Cotton, TR Utterback, MC Hanna, DT Nguyen, DM Saudek, RC Brandon, LD Fine, JL Fritchman, JL Fuhrmann, NSM Geoghagen, CL Gnehm, LA McDonald, KV Small, CM Fraser, HO Smith, JC Venter. *Science 269*:496–512, 1995.

2. AL Delcher, D Harmon, S Kasif, O White, SL Salzberg. *Nucleic Acids Res. 27*: 4636–4641, 1999.
3. SF Altschul, TL Madden, AA Schaffer, J Zhang, Z Zhang, W Miller, DJ Lipman. *Nucleic Acids Res. 25*:3389–3402, 1997.
4. WR Pearson. *Methods Mol. Biol. 132*:185–219, 2000.
5. H Tettelin, KE Nelson, IT Paulsen, TD Read, S Peterson, J Heidelberg, RT DeBoy, DH Haft, RJ Dodson, AS Durkin, M Gwinn, JF Kolonay, WC Nelson, JD Peterson, LA Umayam, O White, SL Salzberg, MR Lewis, D Radune, E Holtzapple, H Khouri, AM Wolf, TR Utterback, CL Hansen, LA McDonald, TV Feldblyum, S Angiuoli, T Dickinson, EK Hickey, IE Holt, BJ Loftus, F Yang, HO Smith, JC Venter, BA Dougherty, DA Morrison, SK Hollingshead, CM Fraser. *Science 293*:498–506, 2001.
6. JC Paton, PW Andrew, GJ Boulnois, TJ Mitchell. *Annu. Rev. Microbiol. 47*:89–115, 1993.
7. RA Alm, LS Ling, DT Moir, BL King, ED Brown, PC Doig, DR Smith, B Noonan, BC Guild, BL deJonge, G Carme, PJ Tummino, A Caruso, M Uria-Nickelsen, DM Mills, C Ives, R Gibson, D Merberg, SD Mills, Q Jiang, DE Taylor, GF Vovis, TJ Trust. *Nature 397*:176–180, 1999.
8. TD Read, RC Brunham, C Shen, SR Gill, JF Heidelberg, O White, EK Hickey, J Peterson, T Utterback, K Berry, S Bass, K Linher, J Weidman, H Khouri, B Craven, C Bowman, R Dodson, M Gwinn, W Nelson, R DeBoy, J Kolonay, G McClarty, SL Salzberg, J Eisen, CM Fraser. *Nucleic Acids Res. 28*:1397–1406, 2000.
9. LA Snyder, SA Butcher, NJ Saunders. *Microbiology 147*:2321–2332, 2001.
10. A Bateman, E Birney, R Durbin, SR Eddy, RD Finn, EL Sonnhammer. *Nucleic Acids Res. 27*:260–262, 1999.
11. DH Haft, BJ Loftus, DL Richardson, F Yang, JA Eisen, IT Paulsen, O White. *Nucleic Acids Res. 29*:41–43, 2001.
12. SF Altschul, EV Koonin. *Trends. Biochem. Sci. 23*:444–447, 1998.
13. RL Tatusov, EV Koonin, DJ Lipman. *Science* 278:631–637, 1997.
14. FS Domingues, P Lackner, A Andreeva, MJ Sippl. *J. Mol. Biol. 297*:1003–1013, 2000.
15. NS Akopyants, A Fradkov, L Diatchenko, JE Hill, PD Siebert, SA Lukyanov, ED Sverdlov, DE Berg. *Proc. Natl. Acad. Sci. USA. 95*:13108–13113, 1998.
16. ML Bogush, TV Velidkodvorskaya, YB Lebedev, LG Nikolaev, SA Lukyanov, AF Fradkov, BK Pliyev, MN Boichenko, GN Usatova, AA Vorobiev, GL Andersen, ED Sverdlov. *Mol. Gen. Genet. 262*:721–729, 1999.
17. JS Anderson, M Mann. *FEBS. Lett. 480*:25–31, 2000.
18. J-C Rain, L Selig, H De Reuse, V Battaglia, C Reverdy, S Simon, G Lenzen, F Petel, J Wojcik, V Schachter, Y Chemama, A Labigne, P Legrain. *Nature 409*: 211–215, 2001.
19. CA Hutchison, SN Peterson, SR Gill, RT Cline, O White, CM Fraser, HO Smith, JC Venter. *Science 286*:2165–2169, 1999.
20. PO Brown, D Botstein. *Nat. Genet. 21(suppl 1)*:33–37, 1999.
21. RD Perry. *Trends Microbiol. 7*:385–388, 1999.
22. SH Lee, SM Butler, A Camilli. *Proc. Natl. Acad. Sci. USA. 98*:6889–6894, 2001.
23. JM Slauch, A Camilli. *Methods Enzymol. 326*:73–97, 2000.

24. M Handfield, LJ Brady, A Progulske-Fox, JD Hillman. *Trends Microbiol 8*:336–339, 2000.
25. H Gmuender, K Kuratli, K DiPadova, CP Gray, W Keck, S Evers. *Genome Res. 11*: 28–42, 2001.
26. A de Saizieu, C Gardes, N Flint, C Wagner, M Kamber, TJ Mitchell, W Keck, KE Amrein, R Lange. *J. Bacteriol 182*:4696–4703, 2000.
27. S Peterson, RT Cline, H Tettelin, V Sharov, DA Morrison. *J. Bacteriol 182*: 6192–6202, 2000.
28. MA Behr, MA Wilson, WP Gill, H Salamon, GK Schoolnik, S Rane, PM Small. *Science 284*:1520–1523, 1999.
29. R Hakenbeck, N Balmelle, B Weber, C Gardes, W Keck, A de Saizieu. *Infect Immun. 69*:2477–2486, 2001.
30. N Salama, K Guillemin, TK McDaniel, G Sherlock, L Tompkins, S Falkow. *Proc. Natl. Acad. Sci. USA. 97*:14668–14673, 2000.
31. JR Fitzgerald, DE Sturdevant, SM Mackie, SR Gill, JM Musser. *Proc. Natl. Acad. Sci. USA. 98*:8821–8826, 2001.
32. C Debouck, PN Goodfellow. *Nat. Genet. 21*:48–50, 1999.
33. DI Papac, Z Shahrokh. *Pharm. Res. 18*:131–145, 2001.
34. MP Washburne, JR Yates III. *Proteomics: A Trends Guide 1*:27–30, 2000.
35. L Anderson, J Seilhamer. *Electrophoresis 18*:533–537, 1997.
36. H Etz, DB Minh, T Henics, A Dryla, B Winkler, C Triska, AP Boyd, W Schmidt, U von Ahsen, M Buschle, S Gill, J Kolonay, C Fraser, A von Gabain, E Nagy, A Meinke. *Proc. Natl. Acad. Sci. USA., 99*:6573–6578, 2002.
37. MR Graham, LM Smoot, BF Lei, JM Musser. *Curr. Opin. Microbiol 4*:65–70, 2001.
38. F Arigoni, F Talabot, M Peitsch, MD Edgerton, E Meldrum, E Allet, R Fish, T Jamotte, ML Curchod, H Loferer. *Nat. Biotechnol* 16:851–856, 1998.
39. SL Chiang, JJ Mekalanos, DW Holden. *Annu. Rev. Microbiol* 53:129–154, 1999.
40. J Horecka, Y Jigami. *Yeast 16*:967–970, 2000.
41. S Fields, O Song. *Nature 340*:245–246, 1989.
42. PL Bartel, JA Roecklein, D SenGupta, S Fields. *Nature Genet. 12*:72–77, 1996.
43. P Uetz, L Giot, G Cagney, TA Mansfield, RS Judson, JR Knight, D Lockshon, V Narayan, M Srinivasan, P Pochart, A Qureshi-Emili, Y Li, B Godwin, D Conover, T Kalbfleisch, G Vijayadamodar, M Yang, M Johnston, S Fields, JM Rothberg. *Nature 403*:623–627, 2000.
44. JC Hu, EK O'Shea, PS Kim, RT Sauer. *Science 250*:1400–1403, 1990.
45. G Karimova, J Pidoux, A Ullmann, D Ladant. *Proc. Natl. Acad. Sci. USA. 95*: 5752–5756, 1998.
46. AL Delcher, S Kasif, RD Fleischmann, J Peterson, O White, SL Salzberg. *Nucleic Acids Res. 27*:2369–2376, 1999.
47. RD Fleischmann, D Alland, J Eisen, L Carpenter, O White, J Peterson, R DeBoy, R Dodson, M Gwinn, D Haft, E Hickey, JF Kolonay, WC Nelson, LA Umayam, M Ermolaeva, S Salzberg, A Delcher, T Utterback, A Mikula, W Bishai, WR Jacobs Jr, JC Venter, CM Fraser. *J. Bacteriol. 184*:5479–5490, 2002.
48. DG Wang, JB Fan, CJ Siao, A Berno, P Young, R Sapolsky, G Ghandour, N Perkins, E Winchester, J Spencer, L Kruglyak, L Stein, L Hsie, T Topaloglou, E Hubbell, E

Robinson, M Mittmann, MS Morris, N Shen, D Kilburn, J Rioux, C Nusbaum, S Rozen, TJ Hudson, R Lipshutz, M Chee, ES Lander. *Science 280*:1077–1082, 1998.
49. MS Bray, E Boerwinkle, PA Doris. *Hum. Mutat. 17*:296–304, 2001.
50. S Sauer, D Lechner, K Berlin, C Plancon, A Heuermann, H Lehrach, IG Gut. *Nucleic Acids Res. 28*:E100, 2000.
51. CDK Bottema, G Sarkar, JD Cassay, S Li, CM Dutton, SS Sommer. *Methods Enzymol. 218*:388–402, 1993.
52. KJ Livak, SJ Flood, J Marmaro, W Giusti, K Deetz. *PCR Methods Appl. 4*:357–362, 1995.
53. SE Brenner. *Nat. Struct. Biol. 7(suppl)*: 967–969, 2000.
54. SK Burley. *Nat. Struct. Biol. 7(suppl)*: 932–934, 2000.
55. SE Brenner, C Chothia, TJ Hubbard. *Curr. Opin. Struct. Biol. 7(3)*:369–376, 1997.
56. AM Edwards, CH Arrowsmith, D Christendat, A Dharamsi, JD Friesen, JF Greenblatt, M Vedadi. *Nat. Struct. Biol. 7(suppl)*: 970–985, 2000.
57. G Grandi. *Trends Biotech. 19*:181–188, 2001.
58. M Pizza, V Scarlato, V Masignani, MM Giuliani, B Arico, M Comanducci, GT Jennings, L Baldi, E Bartolini, B Capecchi, CL Galeotti, E Luzzi, R Manetti, E Marchetti, M Mora, S Nuti, G Ratti, L Santini, S Savino, M Scarselli, E Storni, P Zuo, M Broeker, E Hundt, B Knapp, E Blair, T Mason, H Tettelin, DW Hood, AC Jeffries, NJ Saunders, DM Granoff, JC Venter, ER Moxon, G Grandi, R Rappuoli. *Science 287*:1816–1820, 2000.
59. HM Dockrell, S Brahmbhatt, BD Robertson, S Britton, U Fruth, N Gebre, M Hunegnaw, R Hussain, R Manandhar, L Murillo, MC Pessolani, P Roche, JL Salgado, E Sampaio, F Shahid, JE Thole, DB Young. *Infect. Immun. 68*:5846–5855, 2000.
60. AA Holder. *Proc. Natl. Acad. Sci. USA. 96*:1167–1169, 1999.
61. M Parra, G Hui, AH Johnson, JA Berzofsky, T Roberts, IA Quakyi, DW Taylor. *Infect. Immun. 68*:2685–2691, 2000.
62. YP Shi, SE Hasnain, JB Sacci, BP Holloway, H Fujioka, N Kumar, R Wohlhueter, SL Hoffman, WE Collins, AA Lal. *Proc. Natl. Acad. Sci. USA*. 96: 1615–1620, 1999.
63. DA Relman. *Science* 284:1308–1310, 1999.
64. O Beja, L Aravind, EV Koonin, MT Suzuki, A Hadd, LP Nguyen, SB Jovanovich, CM Gates, RA Feldman, JL Spudich, EN Spudich, EF DeLong. *Science 289*: 1902–1906, 2000.
65. O Beja, MT Suzuki, EV Koonin, L Aravind, A Hadd, LP Nguyen, R Villacorta, M Amjadi, C Garrigues, SB Jovanovich, RA Feldman, EF DeLong. *Environ. Microbiol. 2*:516–529, 2000.
66. KE Nelson, C Weinel, IT Paulson, RJ Dodson, H Hilbert, et al. *Environ. Microbiol. 4*:799–808, 2002.

3

Bioinformatics, Genomics, and Antimicrobial Drug Discovery

Brian A. Dougherty
Bristol-Myers Squibb Pharmaceutical Research Institute,
Wallingford, Connecticut, U.S.A

Daniel B. Davison
Bristol-Myers Squibb Pharmaceutical Research Institute,
Pennington, New Jersey, U.S.A.

INTRODUCTION

The field of genomics has fundamentally changed the way research has been done since the early 1990s and is actively being incorporated into strategies for identifying new drugs and other therapeutics [1]. The emergence of the field of genomics was catalyzed by two crucial factors: the development of automated DNA sequencing [2] and the availability of high-performance computing infrastructure [3]. Upon this foundation, many associated techniques have been developed or improved, such as microarraying, proteomics, and other laboratory and bioinformatic techniques. All of these "genomics technologies" have one aspect in common: the ability to analyze gene function in a high-throughput manner. Antimicrobial drug discovery was one of the first areas to benefit from genomics, due to relatively small genome sizes and, in many cases, systems amenable to genetic manipulation. This chapter describes how bioinformatic and genomic technologies are being leveraged for the discovery of novel antimicrobials.

ANTIMICROBIAL DRUG DISCOVERY

The process of discovering and developing new antimicrobial drugs is outlined in Figure 1. The first step is target identification, which in the genomic era is

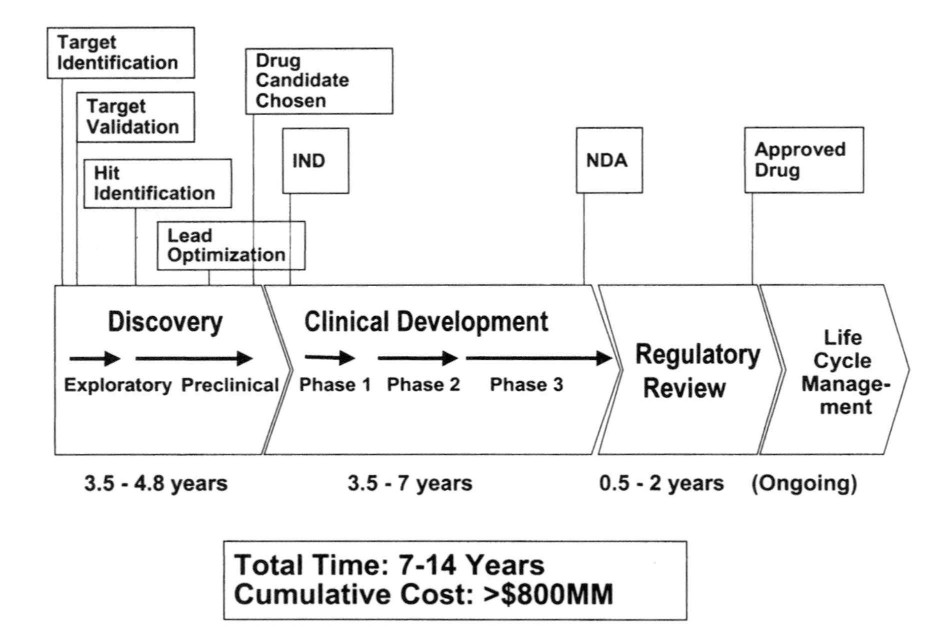

FIGURE 1 Overview of drug discovery and development.

usually done by a bioinformatic filtering process to identify genes specifically conserved in bacteria. The next step is to validate the drug target, usually by confirming that the gene encoding the drug target is essential for viability or, in some cases, essential for virulence [4]. A validated drug target is then used to design an assay that will be used to screen a large deck of small molecule compounds which have been designed by companies from natural products, medicinal chemistry, and combinatorial chemistry. From the screen, a number of "hits" are identified, validated, and prioritized. The hits then progress to "leads" following secondary assays, SAR (structure–activity relationship) studies, and preliminary pharmacology and toxicology studies. At the end of the preclinical process, a drug candidate is chosen and an IND (Investigational New Drug) application is submitted to the Federal Drug Administration. Compound production is scaled up and the drug candidate then goes through clinical trials to determine safety, efficacy, and dosing. Following the successful conclusion of the clinical trials process, an NDA (New Drug Application) is filed with the FDA. Once the drug has been approved by the FDA (about 20% of INDs reach this point), the company markets and sells the drug; new indications are also sought as part of the life-cycle management process. It should be noted that the attrition rate from target to approved drug is high: For an average large pharmaceutical company, it has been estimated that 60 targets are required to result in 20 drug candidates, which, after going through clinical trials, result in 3 new drugs [1]. The process also takes a significant amount of time: In the past, the process from target identification to approved drug has taken an average of 14 years, although to compete more effectively drug companies are striving for an ~8-year timeline.

DNA SEQUENCING

Automated sequencers were developed in the mid-1980s [5] and functionally put to use in the early 1990s for expressed sequence tag (EST) projects [6]. With the infrastructure for high-throughput sequencing, assembly, and data management established, the next major milestone was the publication of the first bacterial genome [7]. This was a critical development for genome sequencing for several reasons. First, the accepted paradigm for sequencing DNA over 50 kb was a top-down, cosmid-by-cosmid approach involving a time- and resource-intensive up-front mapping stage. *Haemophilus influenzae* was sequenced by a whole-genome shotgun method, rapidly accelerating project timelines while still providing very high quality sequence data. Second, the project set a high standard for the publication of future genome sequences: an entire microbial genome without gaps, extensively annotated, and making use of web pages to include the unprecedented volume of data. Finally, the completion of an ~2 Mb segment of DNA suggested the feasibility of a incorporating shotgun sequencing to accelerate the sequencing of the human genome [8]; in fact, the Venter group went on to utilize the whole-genome shotgun technique to completely sequence the ~120 Mb *Drosophila* [9] and ~2910 Mb human [10] genomes.

Whole-genome shotgun sequencing is the most rapid, cost-effective method for sequencing microbial genomes. Since 1995, hundreds of microbial genome projects have been initiated [11] and to date, over 55 microbial genomes have been completed. This number continues to rapidly increase, and lists on the world-wide web should be referred to for the most up-to-date information (see Table 1, "Microbial Genome Project Listings" section). The process of microbial genome sequencing has been reviewed in more detail elsewhere [12,13], but essentially consists of the following steps: [1] constructing a library of randomly sheared DNA, [2] shotgun sequencing of clones for high genome coverage, [3] assembly of the genome into contiguous sequences (contigs), [4]] closure of the gaps between contigs, and [5] gene finding and annotation.

Nearly every step of a genome sequencing project involves computers or bioinformatic tools: for instance, basecalling [e.g., ABI Basecaller, Phred [14], sequence assembly [TIGR Assembler, Phrap [15]], gene finding [GeneMark [16], Glimmer [17], and Critica [18]], and suites of tools for multiple analyses [the GCG package [19] and the genome project management tool MAGPIE [20]].

TABLE 1 Selected Genomics-Related Sites from the World Wide Web

Web site description	Web site URL
Microbial Genome Projects	
TIGR Microbial Database—Complete Genomes	http://www.tigr.org/tdb/mdb/mdbcomplete.html
GOLD—Genomes On Line Database	http://ergo.integratedgenomics.com/GOLD/
NCBI—Completed Genomes	http://www.ncbi.nlm.nih.gov/Entrez/Genome/org.html
Escherichia coli K-12 Genome—Colibri	http://genolist.pasteur.fr/Colibri/
Bacillus subtilis Genome—SubtiList	http://genolist.pasteur.fr/SubtiList/
Bacillus subtilis Genome—Mutant collection	http://locus.jouy.inra.fr/cgi-bin/genmic/madbase/progs/ACCUEIL-MUTANT.pl
Yeast Genome YPD Server	http://www.proteome.com/DB-demo/intro-to-YPD.html
Yeast Genome Deletion Project	http://sequence-www.stanford.edu/group/yeast_deletion_project/
Analysis of Genome Sequence Data	
NCBI – BLAST with Microbial Genomes	http://www.ncbi.nlm.nih.gov/cgi-bin/Entrez/genom_table_cgi
BLAST Programs – Including Psi-BLAST	http://www.ncbi.nlm.nih.gov/BLAST/
BLOCKS Server	http://www.blocks.fhcrc.org/
PROSITE Database	http://www.expasy.ch/prosite/
Pfam Protein Families Database—HMMs	http://www.sanger.ac.uk/Software/Pfam/
COGs (Clusters of Orthologous Groups)	http://www.ncbi.nlm.nih.gov/COG/
SEEBUGS/Microbial GenomeConcordance	http://lion.cabm.rutgers.edu/~bruc/microbes/index.html
CMR (Comprehensive Microbial Resource)	http://www.tigr.org/tigr-scripts/CMR2/CMRHomePage.spl
CLUSTAL Server	http://www.ebi.ac.uk/clustalw/
PAUP Phylogenetic Analysis Using Parsimony	http://www.sinauer.com/Titles/Text/swofford.html
Phylip Phylogeny Inference Package	http://evolution.genetics.washington.edu/phylip.html).
Protein Data Bank—Protein Structures	http://www.rcsb.org/pdb/
EcoCyc: Encyclopedia of *E. coli* Metabolism	http://www.ecocyc.org/
KEGG: Kyoto Encyclopedia of Genes & Genomes	http://www.genome.ad.jp/kegg/
Gene Ontology—Role Classification	http://www.geneontology.org
STRING Server—Recurring Neighboring Genes	http://www.bork.embl-heidelberg.de/STRING/
WIT2 Operon-Pair Spreadsheet	http://wit.mcs.anl.gov/WIT2/CGI/operons.cgi?user=
Microarray Protocol Websites	http://genome.uc.edu/genome/Web_Resources/arrays.html

Moreover, at the core of every successful genome project is a relational database to archive and track everything from raw sequence chromatograms to a series of updated assemblies to properties associated with each gene (e.g., sequence, coordinates on the genome, similarity to other genes, and experimental data) Microbial genome projects can range in effort from low-pass sequencing to rapidly identify the majority of genes in a genome (often seen in industrial settings) to published whole-genome sequencing projects where every gap is closed, representing a firm framework for future experimental work since the genome has become a finite entity.

BIOINFORMATIC ANALYSIS OF GENES

Clearly, bioinformatics is essentially integrated into all aspects of microbial genomics and drug discovery [21]. Some of the major uses of bioinformatic tools are presented below.

Pairwise Similarity Searching

Many programs exist, and some of the most widely used include FASTA [22], the workhorse program BLAST [23], and the more sensitive (and more computationally intensive) Smith–Waterman algorithm [24]. Most genome projects use BLAST to search the genes found within a genome; however, determining what constitutes true signal in the case of weakly similar genes is always a contentious issue. This is especially true for small genes found by genome sequencing projects that are conserved at midrange similarity among different bacteria. In general, about 30–50% of the genes identified in microbial genome sequencing projects are considered unknown, depending on the organism and the annotation criteria used. One effort to enrich for potentially meaningful similarities found among the noise is the use of iterative BLAST searching, known as Psi-BLAST [25,26]. Low-coverage genome projects may, like EST projects, suffer from a higher error rate and programs that take frameshifting into account, such as Framesearch [27], are invaluable.

Motif Searching

Similarity may exist between two proteins that is only apparent when focusing on small regions, such as conserved motifs. For many years researchers have made use of the Prosite database of motifs [28] and the Blocks [29] database. A more recent collection of motifs, the Pfam database [30,31], was constructed using the Hidden Markov model-based HMMER suite of tools. Finally, the Collection of Orthologous Groups (COG) [33] database contains clusters of proteins found throughout sequenced genomes that are potentially related to each other.

Whole-Genome Comparisons

Once whole-genome sequence information began to appear, it became clear that tools designed to analyze entire genomes needed to be designed. The SEEBUGS suite of tools [34] allows users to identify a concordance of genes conserved among selected organisms at any selected similarity level and includes the capability of quickly identifying "neighbors" of selected genes in other genomes and specifically subtracting out selected organisms. This program was designed to address an issue at the heart of antimicrobial drug discovery: identifying potential drug targets (gene products) that are conserved in certain groups of bacteria (e.g., gram-positive and gram-negative) but not found in humans. This allows researchers to build in up front both spectrum and selectivity for novel classes of antibiotics. Another tool that has more recently been developed for whole-genome comparisons is the Comprehensive Microbial Resource (Table 1).

Multiple Sequence Alignments

Alignments of multiple sequences from a diverse set of organisms helps to highlight conserved regions; one of the standard tools for this is the CLUSTAL package [35]. Conserved regions identified by the multiple sequence alignment can then be more carefully analyzed by motif searching techniques to identify additional related genes or by phylogenetic analyses. Also, regions conserved in alignments of linear amino acid sequences can be the focus of three-dimensional modeling and structural similarity searching techniques.

Phylogenetic Analysis

Phylogenetic analysis takes the relationships found among a set of aligned sequences and graphically presents them as a "tree." The most widely used programs are PAUP (Phylogenetic Analysis Using Parsimony; Sinauer Associates, Sunderland, MA) by D. Swofford and the PHYLIP phylogenetic suite of programs by J. Felsenstein and collaborators (available at: evolution.genetics.washington.edu/phylip.html). There are three commonly used methods: distance, parsimony, and maximum likelihood (ML). Each has its benefits and drawbacks. Most molecular biologists use distance methods, while parsimony and ML are generally preferred but require more expertise and computing time. The trees produced can show the evolutionary relationships of the genes under examination (a gene tree) or the species under examination (a species tree). However, it must be noted that these two types of tree are not the same thing and do not provide the same information. Phylogenetic trees can help to classify an unknown reading frame. An excellent review of the use of ML was recently published [36].

Before genome sequences became available, much of the phylogenetic analysis of microbes focused on taxonomic classification of microbes and the establishment of homology among genes. The flood genes from genome sequencing projects made the finding of gene family relationships even more important to understanding the function of gene products and the evolution of micro-organisms. Genome sequence information requires new computational methods that build on the comparison of predicted gene sequence, incorporating other important information such as structures, domain shuffling, expression patterns, and gene adjacency in genomes [37,38].

Structural Analysis

A number of techniques are used to examine relationships among three-dimensional structures of proteins, such as structural homology comparisons of different solved structures or the "threading" of query sequences versus known structures [39]. Some more recent approaches use a combination of techniques, such as Psi-BLAST coupled with structural prediction programs [40]. The field of rational drug design is of particular interest to pharmaceutical and biotechnology companies. Some techniques include virtual screening of decks of compounds or even chemical fragments [41] for hits that fit into an important region of a protein and the use of molecular modeling techniques to optimize a hit into a lead compound [e.g., structure–activity relationship by nuclear magnetic resonance [42]].

Metabolic Pathway Modeling

Genomics technologies such as microarraying, proteomics, and high-throughput biology techniques give vast amounts of data to be analyzed. Powerful methods for providing context to this information involve overlaying the information onto a metabolic pathway map or the assignment of role category information (e.g., cellular role, enzyme class, and gene family). Metabolic pathway modeling has resulted in some excellent knowledge bases, particularly the EcoCyc [Encyclopedia of *E. coli* Genes and Metabolism; [43]] and KEGG [Kyoto Encyclopedia of Genes and Genomes; [44]] databases. One example underscoring the power of having complete genome sequence information in hand for metabolic reconstruction and comparative genomics work is a study examining the presence/absence and conservation of citric acid cycle enzymes across 19 genomes [45]. Role categorization methods have also been developed for organizing microbial genome data, such as the categories developed for many microbial genome projects [modified from Riley [46]] and the Gene Ontology effort that has been used primarily in model organism sequencing projects [47]. There is an ongoing effort to merge these role classification efforts into the Gene Ontology Consortium (available at: www.geneontology.org/#consortium).

Operon Analysis

One feature unique to bacterial genomes is the grouping of genes into operons—coregulated sets of genes that are often related in function. Rigorously defining an operon is a tedious process, especially if done experimentally; however, a number of computer-based prediction methods have been developed [48–51]. One particularly intriguing approach [52] utilizes the gene clustering approach used by others but does so in combination with other data, such as mRNA expression patterns and protein–protein interactions, resulting in a powerful predictive tool.

EXPERIMENTAL TECHNIQUES

A number of genomics technologies exist for the analysis of all genes and/or gene products in a genome. Although these tools are being used for all of the steps in the drug discovery and development process, much of the effort is focused on the use of these technologies for target identification and validation.

Transcriptional Profiling

Microarraying is a high-throughput transcriptional profiling technique that is being applied increasingly to examine differential gene expression in microbial genomes. The technique is based on the hybridization of labeled RNA to miniaturized deposits of DNA on small solid supports. Two widely used formats are synthetic Affymetrix GeneChip arrays, composed of short oligonucletides photolithographically deposited onto silicon wafers [53], and "spotted" microarrays, usually consisting of a gene fragment (several hundred basepairs) deposited onto a glass microscope slide [54]. Basically, RNA derived from various conditions of interest are labeled (one dye in the case of Affymetrix and two dyes for spotted arrays), hybridized to the arrays, and the signal quantitated. After the data have undergone some initial filtering and have been stored into a database, a number of software packages can be utilized for expression analysis. In addition to programs offered by microarray manufacturers, developed in-house by companies, or published by academic groups [55,56], a number of third-party packages are used such as Genespring (Silicon Genetics, Redwood City, CA: available at: www.signetics.com) or Rosetta Resolver (Rosetta Biosoftware, Kirkland, WA; available at: www.rosettabio.com).

There are many recent publications involving expression profiling of microbes; a few that have a drug discovery focus are emphasized here. In one study, the pathogen *Mycobacterium tuberculosis* was treated with the commonly used antibiotic isoniazid, and gene expression changes were monitored using microarrays [57]. Genes observed to change in response to the drug were known to be part of the isoniazid mechanism of action, although some additional genes of unknown function were also affected. In fact, there are many drug mechanism of action studies that have been performed in industry, but few have been published to date. One area of great interest to industrial groups working on antifungal drug discovery is the effect of azoles and other drugs on ergosterol biosynthesis. A recent publication [58] examined the transcriptional profile of azole-treated *Saccharomyces cerevisiae* and compared these patterns to profiles of *S. cerevisiae* strains with knockouts of ergosterol biosynthetic genes. Responsive genes included those in the ergosterol pathway as well as mitochondrial genes, oxidative stress genes, and a few genes of unknown function. Also in this study, an antifungal drug of a novel chemotype was profiled and found to be similar to the azoles, thus underscoring the power of microarray-based drug profiling studies. Another microarray experiment by pharmaceutical researchers focused on two key regulators of virulence in the gram-positive pathogen *Staphylococcus aureus* in an effort to develop treatment against pathogenesis-based targets [59]. Using a custom-designed Affymetrix array, the researchers found genes known or expected to be regulated by *agr* and SarA, but also found known genes not thought to be directly involved in staphylococcal pathogenesis, suggesting that these global transcriptional regulators play a broader role than previously thought.

DNA microarrays are also being used for other genomic experiments, such as whole-genome DNA–DNA hybridizations. Examples of these microarray-based studies that focus on important pathogens include comparisons of BCG vaccine strains to other mycobacterial isolates [60], comparison of isolates of *Streptococcus pneumoniae* to all genes in the sequenced type 4 strain [61], and comparison of divergent lineages of *S. aureus* to the COL genome to identify pathogenicity islands and dispensable genetic material, estimated to be approximately 22% of the genome [62].

Other techniques independent of microarrays can be used to look at differential gene expression in microbes. Techniques that have found some usage include differential display, subtractive hybridization, and quantitative polymerase chain reaction (PCR) [63]. Quantitative reverse-transcription PCR has increasingly been used as the method of choice to validate microarray results. Additionally, the technique has become robust and accurate enough to be used as a transcriptional profiling tool in its own right. For any transcriptional profiling technique using bacteria, a notable experimental distinction is the absence of a usable poly-A tail, requiring the modification of many of the molecular techniques that have been devised using eukaryotic RNA.

Proteomics

Proteomics is the study of the protein complement of the genome. Differential expression can be observed at the protein level for different samples by protein separation followed by staining and comparison of the signal for the separated species. The protein separation technology that has been in use for the past several decades is two-dimensional Sodium dodecyl sulfide–polyacrylamide gel electrophoresis [64], but other separation technologies are being developed, such as multidimensional liquid chromatography [65]. Similarly, Coomassie blue and silver stain are often used today, but, again, newer technologies, such as SYPRO dyes [66], are becoming available. Once proteins are separated and species representing differentially expressed proteins found, the next step is protein identification, which can be done by techniques such as mass spectroscopy (MS/MS), matrix-desisted laser-desorption ionization time-of-flight, and peptide sequencing [67]. Extensive cataloging of microbial proteome components has been performed, such as a study of the *H. influenzae* proteome published 2 years after the genome sequence was determined [68].

Additional High-Throughput Biological Techniques

A number of other experimental techniques are being used to determine the function of genes discovered by microbial genome sequencing projects. One is global gene knockout campaigns using transposon mutagenesis [69], random and directed gene knockouts [70,71], genome footprinting [72], and antisense technology [73]. Additionally, it is also possible to include regulatable promoters, gene fusions, or oligonucleotide tags [74–76] when performing gene knockouts, allowing for large-scale systematic analyses of gene function. Knockouts of every gene in model micro-organisms, such as *Bacillus subtilis* [76] or *Saccharomyces cerevisae* [77], have been performed by the consortia put in place to sequence these genomes. Additionally, techniques that analyze bacterial virulence, such as *in vivo* expression technology and signature-tagged mutagenesis [78], have relied on fairly random approaches; these techniques stand to benefit tremendously from completed genomes and the development of systematic knockouts and associated constructs. A second widely employed technique is the analysis of protein–protein interactions using genetic systems based on the yeast two-hybrid system [79]. Recent genomewide protein–protein interaction studies have been performed for phage [80], bacteria [81], and yeast [82], and dovetailing this technology with oligonucleotide microarrays should bring increased throughout [83]. Finally, advances in screening techniques have lead to a number of "gene-to-screen" technologies, developed by biotechnology companies in response to the explosion in genomic information and the subsequent bottleneck in screening gene products of unknown function [reviewed in [1]]. Many of these techniques are based on thermal denaturation, NMR, and phage display technologies and are broadly applicable to all areas of drug discovery. Much of the work using this technology for antimicrobial drug discovery is proprietary at present, although there are some examples of recently published gene-to-screen work to identify novel antibiotics [84,85].

CONCLUDING REMARKS

The sequencing and publication of over 50 genomes since 1995 should be regarded as a monumental achievement; however, an equally impressive postgenomic effort is now building on that infrastructure, using both computational and laboratory-based high-throughput approaches to understand gene function in the context of microbial physiology. Genomics has helped change the face of microbial research within a decade from a gene-by-gene approach to a more comprehensive, "systems biology" approach [86]. The availability of microbial genome sequences and the means to rapidly make use of it will allow an unprecedented ability to develop therapies to treat infectious diseases, including antibacterial and antifungal compounds [87–89] and new classes of vaccines [90,91]. As the genomes of model eukaryotes and humans are elucidated, a powerful knowledgebase-leveraging microbial pathogenesis and cell biology work should enable the development of a new generation of therapies. These include therapies that target other functions in the microbial cell, such as virulence, or that address infection from the host perspective and will hopefully address the problem of pathogens resistant to antimicrobial therapy [92].

REFERENCES

1. M Cockett, N Dracopoli, E Sigal. *Curr. Opin. Biotechnol. 11*:602–609, 2000.
2. LE Hood, MW Hunkapiller, LM Smith. *Genomics 1*:201–212, 1987.
3. AR Kerlavage, MD Adams, JC Kelley, M Dubnick, J Powell, P Shanmugam, JC Venter, C Fields. Proceedings of the Twenty-Six Annual Hawaii International Conference on System Sciences, Hawaii, 1993.
4. LE Alksne, SJ Projan. (2000). *Curr. Opin. Biotechnol. 11*:625–636, 2000.
5. M Hunkapiller, Skent, M Caruthers, W Dreyer, J Firca, C Giffin, S Horvath, T Hunkapiller, P Tempst, L Hood. *Nature 310*:105–111, 1984.
6. MD Adams, JM Kelley, JD Gocayne, M Dubnick, MH Polymeropoulos, H Xiao, CR Merril, A Wu, B Olde, RF Moreno RF, et al. *Science 252*:1651–1656, 1991.
7. RD Fleischmann, MD Adams, O White, RA Clayton, EF Kirkness, AR Kerlavage, CJ Bult, JF Tomb, BA Dougherty, JM Merrick, K McKenney, G Sutton, W FitzHugh, C Fields, JD Gocayne, J Scott, R Shirley, L Liu, A Glodek, JM Kelley, JF Weldman, CA Phillips, T Spriggs, E Hedblom, MD Cotton, TR Utterback, MC Hanna, DT Nguyen, DM Saudek, RC Brandon, LD Fine, JL Fritchman, JL Fuhmann, NSM Geoghagen, CL Gnehm, LA McDonald, KV Small, CM Fraser, HO Smith, JC Venter. *Science 269*:496–512, 1995.
8. JC Venter, HO Smith, L Hood. *Nature 381*:364–366, 1996.
9. MD Adams, SE Celniker, RA Holt, CA Evans, JD Gocayne, PG Amanatides, SE Scherer, PW Li, RA Hoskins, RF Galle, RA George, SE Lewis, S Richards, M Ashburner, SN Henderson, GG Sutton, JR Wortman, MD Yandell, Q Zhang, LX Chen, RC Brandon, YH Rogers, RG Blazej, M Champe, BD Pfeiffer, KH Wan, C Doyle, EG Baxter, G Helt, CR Nelson, GL Gabor, JF Abrl, A Agbayani, HJ An, C Andrews-Pfannkoch, D Baldwin, RM Ballew, A Basu, J Baxendale, L Bayraktaroglu, EM Beasley, KY Beeson, PV Benos, BP Berman, D Bhandari, S Bolshakov, D Borkova, MR Botchan, J Bouck, P Brokstein, P Brottier, KC Burtis, DA Busam, H Butler, E Cadieu, A Center, I Chandra, JM Cherry, S Cawley, C Dahlke, LB Daven-

port, P Davies, B de Pablos, A Delcher, Z Deng, AD Mays, I Dew, SM Dietz, K Dodson, LE Doup, M Downes, S Dugan-Rocha, BC Dunkov, P Dunn, KJ Durbin, CC Evangelista, C Ferraz, S Ferriera, W Fleischmann, C Fosler, AE Gabrielian, NS Garg, WM Gelbart, K Glasser, A Glodek, F Gong, JH Gorrell, Z Gu, P Guan, M Harris, NL Harris, D Harvey, TJ Heiman, JR Hernandez, J Houck, D Hostin, KA Houston, TJ Howland, MH Wei, C Ibegwam, M Jalali, F Kalush, GH Karpen, Z Ke, JA Kennison, KA Ketchum, BE Kimmel, CD Kodira, C Kraft, S Kravitz, D Kulp, Z Lai, P Lasko, Y Lei, AA Levitsky, J Li, Z Li, Y Liang, X Lin, X Liu, B Mattei, TC McIntosh, MP McLeod, D McPherson, G Merkulov, NV Milshina, C Mobarry, J Morris, A Moshrefi, SM Mount, M Moy, B Murphy, L Murphy, DM Muzny, DL Nelson, DR Nelson, KA Nelson, K Nixon, DR Nusskern, JM Pacleb, M Palazzolo, GS Pittman, S Pan, J Pollard, V Puri, MG Reese, K Reinert, K Remington, RD Saunders, F Scheeler, H Shen, BC Shue, I Siden-Kiamos, M Simpson, MP Skupski, T Smith, E Spier, AC Spradling, M Stapleton, R Strong, E Sun, R Svirskas, C Tector, R Turner, E Venter, AH Wang, X Wang, ZY Wang, DA Wassarman, GM Weinstock, J Weissenbach, SM Williams, T Woodage, KC Worley, D Wu, S Yang, QA Yao, J Ye, RF Yeh, JS Zaveri, M Zhan, G Zhang, Q Zhao, L Zheng, XH Zheng, FN Zhong, W Zhong, X Zhou, S Zhu, X Zhu, HO Smith, RA Gibbs, EW Myers, GM Rubin, JC Venter. *Science 287*:2185–2195, 2000.

10. JC Venter, MD Adams, EW Myers, PW Li, RJ Mural, GG Sutton, HO Smith, M Yandell, CA Evans, RA Holt, JD Gocayne, P Amanatides, RM Ballew, DH Huson, JR Wortman, Q Zhang, CD Kodira, XH Zheng, L Chen, M Skupski, G Subramanian, PD Thomas, J Zhang, GL Gabor Miklos, C Nelson, S Broder, AG Clark, J Nadeau, VA McKusick, N Zinden, AJ Levine, RJ Roberts, M Simon, C Slayman, M Hunkapiller, R Bolanos, A Delcher, I Dew, D Fasulo, M Flanigan, L Florea, A Halpern, S Hannenhalli, S Kravitz, S Levy, C Mobarry, K Reinert, K Remington, J Abu-Threideh, E Beasley, K Biddick, V Bonazzi, R Brandon, M Cargill, I Chandramouliswaran, R Charlab, K Chaturvedi, Z Deng, V Di Francesco, P Dunn, K Eilbeck, C Eyangelista, AE Gabrielian, W Gan, W Ge, F Gong, Z Gu, P Guan, TJ Heiman, ME Higgins, RR Ji, Z Ke, KA Ketchum, Z Lai, Y Lei, Z Li, J Li, Y Liang, X Lin, F Lu, GV Merkulov, N Milshina, HM Moore, AK Naik, VA Narayan, B Neelam, D Nusskern, DB Rusch, S Salzberg, W Shao, B Shue, J Sun, Z Wang, A Wang, X Wang, J Wang, M Wei, R Wides, C Xiao, C Yan, A Yao, J Ye, M Zhan, W Zhang, H Zhang, Q Zhao, L Zheng, F Zhong, W Zhong, S Zhu, S Zhao, D Gilbert, S Baumhueter, G Spier, C Carter, A Cravchik, T Woodage, F Ali, H An, A Awe, D Baldwin, H Baden, M Barnstead, I Barrow, K Beeson, D Busam, A Carver, A Center, ML Cheng, L Curry, S Danaher, L Davenport, R Desilets, S Dietz, K Dodson, L Doup, S Ferriera, N Garg, A Gluecksmann, B Hart, J Haynes, C Haynes, C Heiner, S Hladun, D Hostin, J Houck, T Howland, C Ibegwam, J Johnson, F Kalush, L Kline, S Koduru, A Love, F Mann, D May, S McCawley, T McIntosh, I McMullen, M Moy, L Moy, B Murphy, K Nelson, C Pfannkoch, E Pratts, V Puri, H Qureshi, M Reardon, R Rodriguez, YH Rogers, D Romblad, B Ruhfel, R Scott, C Sitter, M Smallwood, E Stewart, R Strong, E Suh, R Thomas, NN Tint, S Tse, C Vech, G Wang, J Wetter, S Williams, M Williams, S Windsor, E Winn-Deen, K Wolfe, J Zaveri, K Zaveri, JF Abril, R Guigo, MJ, Campbell, KV Sjolander, B Karlak, A

Kejariwal, H Mi, B Lazareva, T Hatton, A Narechania, K Diemer, A Muruganujan, N Guo, S Sato, V Bafna, S Istrail, R Lippert, R Schwartz, B Walenz, S Yooseph, D Allen, A Basu, J Baxendale, L Blick, M Caminha, J Carnes-Stine, P Caulk, YH Chiang, M Coyne, C Dahlke, A Mays, M Dombroski, M Donnelly, D Ely, S Esparham, C Fosler, H Gire, S Glanowski, K Glasser, A Glodek, M Gorokhov, K Graham, B Grobman, M Harris, J Heil, S Henderson, J Hoover, D Jennings, C Jordan, J Jordan, J Kasha, L Kagan, C Kraft, A Levitsky, M Lewis, X Liu, J Lopez, D Ma, W Majoros, J McDaniel, S Murphy, M Newman, T Nguyen, N Nguyen, M Nodell, S Pan, J Peck, M Peterson, W Rowe, R Sanders, J Scott, M Simpson, T Smith, A Sprague, T Stockwell, R Turner, E Venter, M Wang, M Wen, D Wu, M Wu, A Xia, A Zandieh, X Zhu. *Science 291*:1304–1351, 2001.

11. KE Nelson, IT Paulsen, JF Heidelberg, CM Fraser. *Nat. Biotechnol. 18*:1049–1054, 2000.
12. CM Fraser, RD Fleischmann. *Electrophoresis 18*:1207–1216, 1997.
13. BA Dougherty. In: RH Baltz, GD Hegeman, PL Skatrud, ed. *Developments in Industrial Microbiology—GMBIM*. Fairfax, VA: Society for Industrial Microbiology, 1997, pp 9–12.
14. B Ewing, L Hillier, MC Wendl, P Green. *Genome Res. 8*:175–185, 1998.
15. B Ewing, P Green. *Genome Res. 8*:186–194, 1998.
16. K Isono, JD McIninch, M Borodovsky. *DNA Res. 1*:263–269, 1994.
17. SL Salzberg, AL Delcher, S Kasif, O White. *Nucleic Acids Res. 26*:544–548, 1998.
18. JH Badger, GJ Olsen. *Mol. Biol. Evol. 16*:512–524, 1999.
19. DD Womble. *Methods Mol. Biol. 132*:3–22, 2000.
20. T Gaasterland, CW Sensen. *Trends Genet. 12*:76–78, 1996.
21. DB Searls. *Drug Discov. Today 5*:135–143.
22. WR Pearson, DJ Lipman. *Proc. Natl. Acad. Sci. USA 85*:2444–2448, 1988.
23. SF Altschul, TL Madden, AA Schaffer, J Zhang, Z Zhang, W Miller, DJ Lipman. *Nucleic Acids Res. 35*:3389–3402, 1997.
24. TF Smith, MS Waterman, WM Fitch. *J. Mol. Evol. 18*:38–46, 1981.
25. SF Altschul, EV Koonin. *Trends Biochem. Sci. 23*:444–447, 1998.
26. KS Kakarova, L Aravind, EV Koonin. *Protein Sci. 18*:1714–1719, 1999.
27. Edelman *et al.* A rigorous program for searching protein databases with nucleic acid queries. Genome Sequence and Analysis Conference, Hilton Head, South Carolina, 1995.
28. K Hofmann, P Bucher, L Falquet, A Bairoch. *Nucleic Acids Res 27*:215–219, 1999.
29. S Henikoff, JG Henikoff. *Proc. Natl. Acad. Sci. USA 89*:10915–10919, 1992.
30. EL Sonnhammer, SR Eddy, R Durbin. *Proteins 28*:405–420, 1997.
31. A Bateman, E Birney, R Durbin, SR Eddy, KL Howe, EL Sonnhammer. *Nucleic Acids Res. 28*:263–266, 2000.
32. EL Sonnhammer, SR Eddy, E Birney, A Bateman, R Durbin *Nucleic Acids Res. 26*: 320–322, 1998.
33. RL Tatusov, EV Koonin, DJ Lipman. *Science 278*:631–637, 1997.
34. RE Bruccoleri, TJ Dougherty, DB Davison. *Nucleic Acids Res. 26*:4482–4486, 1998.
35. DG Higgins, JD Thompson, TJ Gibson. *Methods Enzymol. 266*:383–402, 1996.
36. S Whelan, P Lio, N Nick Goldman. *Trends Genet. 17*:262–272, 2001.

37. JW Thornton, R DeSalle. *Ann. Rev. Genom. Hum. Genet. 1*:41–73, 2000.
38. MY Galperin, EV Koonin. *Nat. Biotechnol. 18*:609–613, 2000.
39. D Baker, A Sali. *Science 294*:93–96, 2001.
40. RL Dunbrack Jr. *Proteins 37(S3)*:81–87, 1999.
41. PJ Hajduk, A Gomtsyan, S Didomenico, M Cowart, EK Bayburt, L Solomon, J Severin, R Smith, K Walter, TF Holzman, A Stewart, S McGaraughty, MF Jarvis, EA Kowaluk, SW Fesik. *J. Med. Chem. 43*:4781–4786, 2000.
42. SB Shuker, PJ Hajduk, RP Meadows, SW Fesik. *Science 274*:1531–1534, 1996.
43. PD Karp, M Riley, SM Paley, A Pelligrini-Toole. *Nucleic Acids Res. 24*:32–39, 1996.
44. H Bono, H Ogata, S Goto, M Kanehisa. *Genome Res. 8*:203–210, 1998.
45. MA Huynen, T Dandekar, P Bork. *Trends Microbiol. 7*:281–291, 1999.
46. M Riley. *Microbiol. Rev. 57*:862–952, 1993.
47. M Ashburner, CA Ball, JA Blake, D Botstein, H Butler, JM Cherry, AP Davis, K Dolinski, SS Dwight, JT Eppig, MA Harris, DP Hill, L Issel-Tarver, A Kasarskis, S Lewis, JC Matese, JE Richardson, M Ringwald, GM Rubin, G Sherlock. *Nat. Genet. 25*:25–29, 2000.
48. R Overbeek, M Fonstein, M D'Souza, GD Pusch, N Maltsev. *Proc. Natl. Acad. Sci. USA 96*:2896–2901, 1999.
49. B Snel, G Lehmann, P Bork. MA Huynen. *Nucleic Acids Res. 28*:3442–3444, 2000.
50. MD Ermolaeva, O White, SL Salzberg. *Nucleic Acids Res. 29*:1216–1221, 2001.
51. AJ Enright, I Iliopoulos, NC Kyrpides, CA Ouzounis. *Nature 402*:86–90, 1999.
52. EM Marcotte, M Pellegrini, MJ Thompson, TO Yeates, D Eisenberg. *Nature 402*: 83–86, 1999.
53. AC Pease, D Solas, EJ Sullivan, MT Cronin, CP Holmes, SP Fodor. *Proc. Natl. Acad. Sci. USA 91*:5022–5026, 1994.
54. M Schena, D Shalon, RW Davis, PO Brown. *Science 270*:467–470, 1995.
55. MB Eisen, PT Spellman, PO Brown, D Botstein. *Proc. Natl. Acad. Sci. USA 95*: 14863–14868, 1998.
56. P Tamayo, D Slonim, J Mesirov, Q Zhu, S Kitareewan, E Dmitrovsky, ES Lander, TR Golub. *Proc. Natl. Acad. Sci. USA 96*:2907–2912, 1999.
57. M Wilson, J DeRisi, HH Kristensen, P Imboden, S Rane, PO Brown, GK Schoolnik. *Proc. Natl. Acad. Sci. USA 96*:12833–12838, 1999.
58. GF Bammert, JM Fostel. *Antimicrob. Agents Chemother. 44*:1255–1265, 2000.
59. PM Dunman, E Murphy, S Haney, D Palacios, G Tucker-Kellogg, S Wu, EL Brown, RJ Zagursky, D Shaes, SJ Projan. *J. Bacteriol. 183*:7341–7353, 2001.
60. MA Behr, MA Wilson, WP Gill, H Salamon, GK Schoolnik, S Rane, PM Small. *Science 284*:1520–1523, 1999.
61. H Tettelin, KE Nelson, IT Paulsen, JA Eisen, TD Read, S Peterson, J Heidelberg, RT DeBoy, DH Haft, RJ Dodson, AS Durkin, M Gwinn, JF Kolonay, WC Nelson, JD Peterson, LA Umayam, O White, SL Salzberg, MR Lewis, D Radune, E Holtzapple, H Khouri, AM Wolf, TR Utterback, CL Hansen, LA McDonald, TV Feldblyum, S Angiuoli, T Dickinson, EK Hickey, IE Holt, BJ Loftus, F Yang, HO Smith, JC Venter, BA Dougherty, DA Morrison, SK Hollingshead, CM Fraser. *Science 293*: 498–506, 2001.

62. JR Fitzgerald, DE Sturdevant, SM Mackie, SR Gill, JM Musser. *Proc. Natl. Acad. Sci. USA 98*:8821–8826, 2001.
63. WM Freeman, SJ Walker, KE Vrana. *Biotechniques 26*:112–122, 124–125, 1999.
64. RA VanBogelen, EE Schiller, JD Thomas, FC Neidhardt. *Electrophoresis 20*: 2149–2159, 1999.
65. GJ Opiteck, SM Ramirez, JW Jorgenson, MA Moseley 3rd. *Anal. Biochem. 258*: 349–361, 1998.
66. JX Yan, RA Harry, C Spibey, MJ Dunn. *Electrophoresis 21*:3657–3665, 2000.
67. JR Yates 3rd. *Trends Genet. 16*:5–8, 2000.
68. AJ Link, LG Hays, EB Carmack, JR Yates 3rd. *Electrophoresis 18*:1314–1334, 1997.
69. BJ Akerley, EJ Rubin, A Camilli, DJ Lampe, HM Robertson, JJ Mekalanos. *Proc. Natl. Acad. Sci. USA 95*:8927–8932, 1998.
70. MS Lee, BA Dougherty, AC Madeo, DA Morrison. *Appl. Environ. Microbiol. 65*: 1883–1890, 1998.
71. F Arigoni, F Talabot, M Peitsch, MD Edgerton, E Meldrum, E Allet, R Fish, T Jamotte, ML Curchod, H Loferer. *Nat. Biotechnol. 16*:851–856, 1998.
72. V Smith, D Botstein, PO Brown. *Proc. Natl. Acad. Sci. USA 92*: 6479–6483, 1995.
73. Y Ji, B Zhang, SF Van Horn, P Warren, G Woodnutt, MK Burnham, M Rosenberg. *Science 293*:2266–2269, 2001.
74. N Ogasawara. *Res. Microbiol. 151*:129–134, 2000.
75. B Dujon. *Electrophoresis 19*:617–624, 1998.
76. V Vagner, E Dervyn, SD Ehrlich. *Microbiol. 144(11)*:3097–3104, 1998.
77. DD Shoemaker, DA Lashkari, D Morris, M Mittmann, RW Davis. *Nat. Genet. 14*: 450–456, 1996.
78. SL Chiang, JJ Mekalanos, DW Holden. *Ann. Rev. Microbiol. 53*:129–154, 1999.
79. S Fields, O Song. *Nature 340*:245–246, 1989.
80. PL Bartel, JA Roecklein, D SenGupta, S Fields. *Nat. Genet. 12*:72–77, 1996.
81. JC Rain, L Selig, H De Reuse, V Battaglia, C Reverdy, S Simon, G Lenzen, F Petel, J Wojcik, V Schachter, Y Chemama, A Labigne, P Legrain. *Nature 409*:211–215, 2001.
82. B Schwikowski, P Uetz, S Fields. *Nat. Biotechnol. 18*:1257–1261, 2000.
83. RJ Cho, M Fromont-Racine, L Wodicka, B Feierbach, T Stearns, P Legrain, DJ Lockhart, RW Davis. *Proc. Natl. Acad. Sci. USA 95*:3752–3757, 1998.
84. R Hyde-DeRuyscher, LA Paige, DJ Christensen, N Hyde-DeRuyscher, A Lim A, ZL Fredericks, J Kranz, P Gallant, J Zhang, SM Rocklage, DM Fowlkes, PA Wendler, PT Hamilton PT. *Chem. Biol. 7*:17–25, 2000.
85. DJ Christensen, EB Gottlin, RE Benson, PT Hamilton. *Drug Discov. Today 6*: 721–727, 2001.
86. T Ideker, V Thorsson, JA Ranish, R Christmas, J Buhler, JK Eng, R Bumgarner, DR Goodlett, R Aebersold, L Hood. *Science 292*:929–934, 2001.
87. TD Read, SR Gill, H Tettelin, BA Dougherty. *Drug Discov. Today 6*:887–892, 2001.
88. AE Allsop. *Curr. Opin. Microbiol. 1*:530–534, 1998.
89. CP Gray, W Keck. *Cell. Mol. Life Sci. 56*:779–787, 1999.
90. M Pizza, V Scarlato, V Masignani, MM Giuliani, B Arico, M Comanducci, GT Jennings, L Baldi, E Bartolini, B Capecchi, CL Galeotti, E Luzzi, R Manetti, E

Marchetti, M Mora, S Nuti, G Ratti, L Santini, S Savino, M Scarselli, E Storni, P Zuo, M Broeker, E Hundt, B Knapp, E Blair, T Mason, H Tettelin, DW Hood, AC Jeffries C, NJ Saunders, DM Granoff, JC Venter, ER Moxon, RG Grandi, R Rappuoli. *Science 287*:1816–1820, 2000.

91. TM Wizemann, JH Heinrichs, JE Adamou, AL Erwin, C Kunsch, GH Choi, SC Barash, CA Rosen, HR Masure, E Tuomanen, A Gayle, YA Brewah, W Walsh, P Barren, R Lathigra, M Hanson, S Langermann, S Johnson, S Koenig. *Infect. Immun. 69*:1593–1598, 2001.
92. D Mazel, J Davies. *J. Cell Mol. Life Sci. 56*:742–754, 1999.

4

The Pathway Tools Software and Its Role in Antimicrobial Drug Discovery

Peter D. Karp
SRI International, Menlo Park, California, U.S.A.

INTRODUCTION

The Pathway Tools software developed by SRI International is a powerful environment for genomics-based antimicrobial drug discovery. With approximately 70 complete microbial genomes now in the public domain, and many others in the private domain, bioinformatics tools for analysis of microbial genomes at a global level are important at all phases of the drug discovery process. The Pathway Tools use biochemical pathways as an organizing framework for genomic data. The pathway framework is an information-reduction device: It transforms the genome from a list of thousands of gene products to a cognitively more manageable list of hundreds of pathways. More important, it transforms our conception of the genome from a list whose members are arranged in unknown relationships to a network whose members are related through explicit causal links.

The Pathway Tools software revolves around the concept of a Pathway/Genome Database (PGDB). A PGDB describes the genome of an organism [its chromosome(s), plasmid(s), genes, and genome sequence], the product of each gene, the biochemical reaction(s) catalyzed by each gene product, the substrates of each reaction, and the organization of those reactions into pathways.

The power of the Pathway Tools is derived from both its database schema and its software components. Both were originally developed for the EcoCyc project [1,2] (available at: http://ecocyc.org/). The database schema (ontology) has been carefully designed to encode the complexities of genome and pathway data. Once a genome is encoded within the Pathway Tools schema, new types of complex analyses become possible because so many important semantic relationships have been described in a structured fashion. The Pathway Tools software components support several different tasks, including querying and visualization, analysis, interactive editing, and Web publishing.

The PGDB constitutes an evolving knowledge model of the organism that can be published on an organization's internal Web site using the Pathway Tools. The Pathway Tools provides a comprehensive and unique collection of query and visualization tools that can be used to determine the current state of knowledge regarding the organism: What function is attributed to a given gene product? What role does it play in a given pathway? How are the genes in that pathway distributed in the genome? To ensure that the PGDB reflects the most up-to-date knowledge about an organism, the Pathway Tools provides a rich set of interactive software tools for curation of a PGDB.

This chapter provides an example of how a drug discovery team would use the Pathway Tools in a genomics-based drug discovery project.

OVERVIEW OF PATHWAY TOOLS COMPONENTS

The Pathway Tools software components, and their primary functions, are as follows.

PathoLogic

Given an annotated genome of a pathogenic micro-organism as its input, PathoLogic predicts the metabolic pathways of the organism, providing new global insights about its biochemistry.

PathoLogic creates a new PGDB describing the genome and the pathways of the organism.

PathoLogic can generate reports that summarize the evidence for the presence of each predicted metabolic pathway.

Pathway/Genome Editors

The Editors are a set of forms-based editing tools that are specially designed to update the different data types within a PGDB.

The Editors are used to alter the content of a PGDB to reflect the evolving knowledge of the biochemical machinery of the organism, such as modifying the functional annotations of genes, the assignment of gene products to pathways, and the connectivity of reactions assigned to a pathway.

Pathway/Genome Navigator

Scientists employ the Navigator to query and visualize the contents of a PGDB.

Visualization tools include a chromosome browser, pathway viewer, and displays of operons and regulons.

The Navigator can run in a WWW mode to publish a PGDB on the Internet or an intranet.

The Navigator can perform comparisons of the entire metabolic networks of multiple organisms.

The Navigator can paint gene-expression or protein-expression data sets on a diagram of the full metabolic network of an organism.

We next consider each of the three software components in more detail.

PATHOLOGIC

Use of the Pathway Tools begins with an annotated genome of a pathogenic microorganism, meaning a complete genome sequence (closed or gapped) for which bioinformatics analyses have already identified the locations of likely coding regions and have predicted the functions of these genes. The genome can be provided in the form of a Genbank file or in a PathoLogic-specific format that can accommodate more information than can the Genbank format. If the Genbank format is used, it is important that the input file adhere closely to the Genbank standard, which, unfortunately, many completed genomes fail to do [3]. Figure 1a shows a sample portion of a Genbank file input to PathoLogic; Figure 1b shows the same information in PathoLogic Format.

In brief, PathoLogic imports the genes and proteins described by the input files into a new PGDB that is structured using the Pathway Tools schema and then matches the enzymes listed in the annotated genome against the enzymes required by every pathway in the MetaCyc DB [1]. Those pathways with significant matches are imported into the new PGDB. More precisely, the algorithm followed by PathoLogic is as follows.

1. Initialize the schema of the new PGDB by replicating the MetaCyc schema.
2. Create a DB object for each chromosome or plasmid in the input file(s). Then create a DB object for each gene in the input file(s) and for the gene product of each gene.
3. Attempt to determine the reaction catalyzed (if any) by each gene product in the organism. If an EC number was assigned to the gene product in the input file, then rely on the EC number. Otherwise, match the name of each gene product against the extensive dictionary of enzyme names within MetaCyc.

(a)

```
gene            422054..423490
                /note="CT370"
                /gene="aroE"
CDS             422054..423490
                /gene="aroE"
                /codon_start=1
                /transl_table=11
                /product="Shikimate 5-Dehydrogenase"
                /db_xref="PID:g3328794"
```

(b)

```
ID              CT370
NAME            aroE
STARTBASE       422054
ENDBASE         423490
PRODUCT         shikimate 5-dehydrogenase
DBLINK          PID:g3328794
PRODUCT-TYPE    P
EC              1.1.1.25
```

Figure 1 A fragment of an input file to PathoLogic for a single gene. (a) Genbank format. (b) PathoLogic format.

4. Match the list of reactions now known to be catalyzed by the organism from step [3] against all MetaCyc pathways. For pathways with significant numbers of matches, import the pathway and its associated reactions and substrates from MetaCyc into the new PGDB.

The diversity of pathways in MetaCyc influences the power of the PathoLogic pathway predictions: The broader the space of pathways in MetaCyc, the larger the range of pathways that PathoLogic can recognize in a new organism. MetaCyc version 7.0 (February 2003) contains 467 pathways from 174 different organisms. Although the majority of MetaCyc pathways were experimentally elucidated in bacteria, approximately 50 MetaCyc pathways were elucidated in plants or animals, and many of the bacterial pathways are likely to be shared by plants and animals. Table 1 shows the number of pathways present in MetaCyc from the most frequently occurring species. Some MetaCyc pathways are labeled as occurring in more than one species.

PathoLogic can generate reports that summarize the amount of evidence supporting each pathway predicted to be present in the new PGDB and that list the “pathway holes”—the enzymes missing from each predicted pathway.

TABLE 1 The Number of MetaCyc Pathways Marked as Occurring in Specified Species

Pathways	Organism
173	*E. coli*
35	*Salmonella typhimurium*
31	*H. sapiens*
20	*Sulfolobus solfataricus*
18	*B. subtilis*
18	*Soybean*
17	*Pseudomonas*
15	*Hm. influenzae*
12	*Mycoplasma capricolum*
7	*Pseudomonas putida*
7	*Mycoplasma pneumoniae*
6	*Ascomycotina*
5	*Rhizobiaceae*
5	*S. cerevisiae*
4	*Clostridium*
4	*Thauera aromatica*
4	*Pseudomonas aeruginosa*
4	*Thermotoga maritima*
4	*Thermotoga maritime*
4	*Rhodococcus*
4	*Klebsiella pneumoniae*
3	*Pseudomonadacea*
3	*Pseudomonas sp.*
3	*Neisseriaceae*
3	*Klebsiella aerogenes*
3	*Rattus norvegicus*
3	*Archaebacteria*
3	*Methanosarcina barkeri*
3	*Archaea*
2	*Pseudomonas aureofaciens*
2	*Brevibacterium*
2	*Arthrobacter globiformis*
2	*Oryctolagus cuniculus*
2	*Mammalia*
2	*Lactobacillaceae*
2	*Pseudomonas acidovorans*
2	*Pseudomonas putida ATCC 17453*
2	*Sphingomonas sp RW1*
2	*Azotobacter beijerinckii*
2	*Acinetobacter*

(*continued*)

TABLE 1 Continued

Pathways	Organism
2	*Comamonas testosteroni T-2*
2	*Aerobacter aerogenes*
2	*Methanosarcina thermophila*
2	*Methanogens*
2	*Synechocystis sp. strain PCC 6803*
2	*Sinorhizobium meliloti*
2	*Halobacterium salinarium*
2	*Thiobacillus ferrooxidans*

PATHWAY/GENOME NAVIGATOR

The Navigator software provides a scientist with the ability to interrogate a PGDB and visualize the results of a query in an intuitive, graphical fashion. It also provides analysis operations, such as whole-metabolic-map comparisons across multiple organisms. The Navigator functionality is available through both X-windows and the World Wide Web (WWW).

The most advanced capability within the Navigator is its ability to generate a diagram of the full metabolic network of the organism, called the Overview Diagram (see Figure 2). The Overview is a device for visualizing global relationships within the metabolic network of the organism. The user can employ an extensive menu of query operations to interrogate the PGDB and visualize the results on the Overview. Queries supported by the Overview include the following:

- Highlight all reaction steps shared with other organisms for which PGDBs are available or unique to the current organism with respect to other organisms
- Highlight all reaction steps for which the enzyme is activated or inhibited by a specified metabolite or for which the enzyme is located in a specified cellular compartment
- Highlight all reaction steps for which the reaction is used in multiple metabolic pathways (in which case, knocking out that enzyme will knock out multiple pathways), or highlight those reactions for which the cell has multiple isozymes that catalyze the reaction
- Paint a protein-expression or gene-expression data set onto the Overview by painting each reaction step in the diagram with a color that reflects the expression level of the enzyme that catalyzes that reaction or of its gene

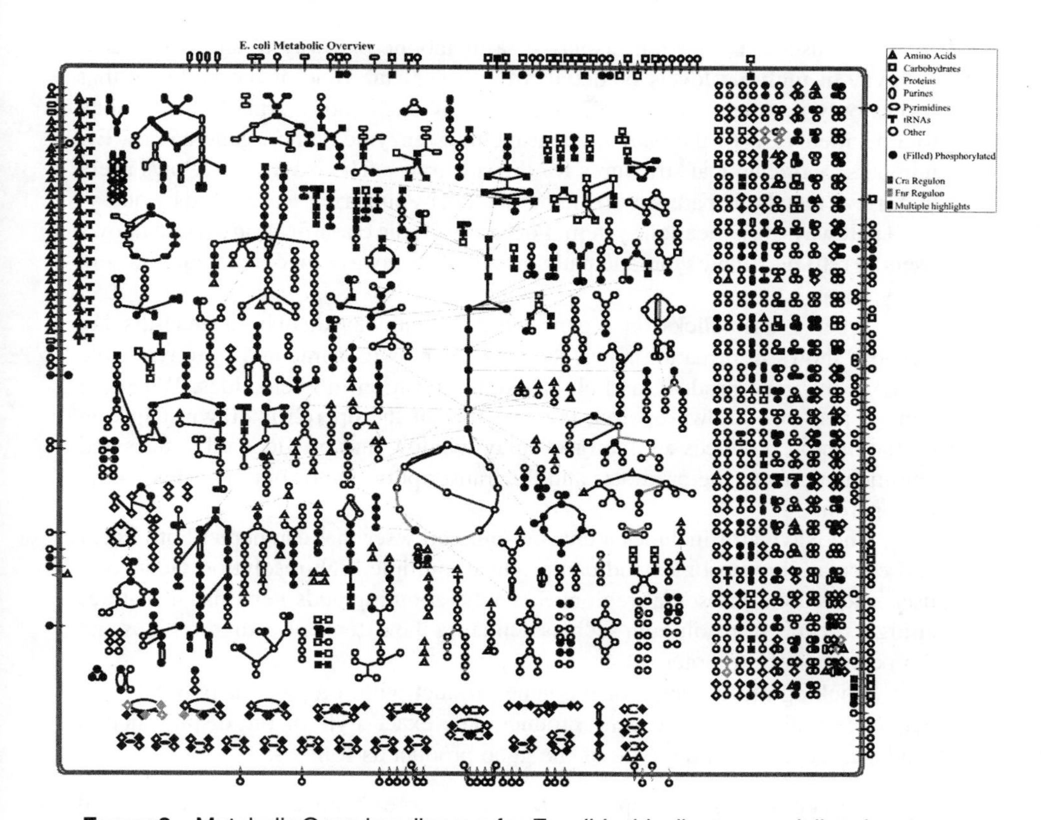

FIGURE 2 Metabolic Overview diagram for *E. coli.* In this diagram each line denotes a single metabolic or transport reaction, and each node denotes a single metabolite. Metabolite shapes encode their chemical class; for example, triangles represent amino acids and squares represent carbohydrates (see legend at top right).

- Highlight all reaction steps for which the enzymes catalyzing the reaction are all controlled by a user-specified transcription factor
- Highlight specific entities in the diagram by name or substring search, such as the name of a pathway, enzyme, or metabolite

The user can also select an entity within the Overview for closer inspection, such as to display a given pathway, enzyme, or metabolite in a window of its own. Each of the Navigator display windows for individual entities (such as a pathway) allows the user to display other related entities by clicking on them. For example, the Navigator pathway window displays all of the reactions, enzymes, and metabolites within a pathway, all of which are clickable.

The user can alter the display of a metabolic pathway to show the same pathway at multiple levels of detail, from a skeletal view of the pathway that shows only the main substrates at the pathway ends and internal branch points to a detailed view of the pathway that includes enzyme names, gene names, EC numbers, and chemical structures for the substrates. The Navigator allows users to retrieve metabolic pathways by name search, by substring search, and by querying a pathway-classification system. For example, the classification system allows users to retrieve all biosynthetic pathways or all pathways involved in amino acid biosynthesis.

When a user clicks on a gene within a pathway display, a gene-display window shows information about the gene such as its name and synonyms, the name of the gene product, and chromosomal position in nucleotides. When the gene is part of a known operon, the structure of the operon is drawn. Clicking on the operon produces an operon-display window that provides references and commentary about the promoter and the transcription-factor binding sites within the operon.

The Navigator includes a chromosome browser that can depict both linear and circular chromosomes and can produce multiple high-resolution views of a user-selected chromosomal region. Semantic zooming adds new visual features at higher levels of resolution, such as depicting the extents of coding regions and the positions of promoters.

Clicking on the name of the gene product within a gene-display window generates a display of the gene product. The exact style of the display that is produced depends on the type of the gene product as follows:

- For enzymes, the display shows one or more chemical reactions catalyzed by the enzyme and displays information about each reaction, including the cofactors, activators, and inhibitors that modulate the activity of the enzyme
- For transporters, the display provides a graphic depiction of the transport activity, indicating its energy-coupling mechanism (e.g., ATP-driven transport versus symport) and the transported substrate
- For transcription factors, the display depicts all of the operons that contain a binding site for that transcription factor (the regulon of the transcription factor)

PATHWAY TOOLS EDITORS

The accuracy of global genome analyses such as comparisons of the metabolic networks of multiple organisms, and interpretation of expression data, depends intimately on the accuracy of the genome annotation. In a typical microbial genome, 30–40% of the genes will have no predicted function. Of those genes that do have predicted functions, some number of those functional predictions will be incorrect [Brenner estimates a 7% error rate [4]]. We postulate that another source of errors in genome annotation is underannotation of multifunctional proteins. Most complete genomes have few proteins that are annotated with more than one predicted function, yet of the 607 enzymes in the EcoCyc DB, 100 are multifunctional [5]. If this frequency of multifunctional proteins is representative of the full genomes of other organisms, *many* protein functions are being missed by genome analysis.

The point of this discussion is that the initial annotation of most genomes is incomplete and error prone and that as more knowledge about the genome is gained, it is important to update the PGDB to reflect the current state of knowledge regarding the genome. A PGDB can serve as a vehicle for recording the current best knowledge of the genome annotation and the rationale for the annotation and can be used to disseminate that knowledge within an organization through the WWW Navigator. The rationale for the genome annotation can be encoded both through PGDB fields that indicate whether a given gene function was determined computationally or experimentally and through *history notes* within a gene that allow users to record time-stamped, name-stamped comments describing why annotations were changed.

The purpose of the Pathway/Genome Editors is to allow users to efficiently update the genome and pathway annotations within a PGDB to reflect the evolving understanding of the organism. The Editors are a collection of interactive forms such as that shown in Figure 3. Each form is designed to support the editing of one of the datatypes within a PGDB. The editing tools include the following:

- A pathway editor for creating new metabolic pathways and modifying existing pathways
- A protein editor for creating and modifying enzyme and transporter descriptions
- A gene editor for creating and modifying genes
- A compound editor for creating and modifying small molecules within a PGDB
- A transcription-unit editor for creating and modifying descriptions of transcription units and transcription factors

THE OCELOT OBJECT DATABASE SYSTEM

Data management services for the Pathway Tools are provided by an object-oriented database system developed called Ocelot [6]. Ocelot combines the expressive power of the frame knowledge representation systems [7] developed within the Artificial Intelligence (AI) community with the scalability, multiuser access capabilities, and robust operation of relational database systems.

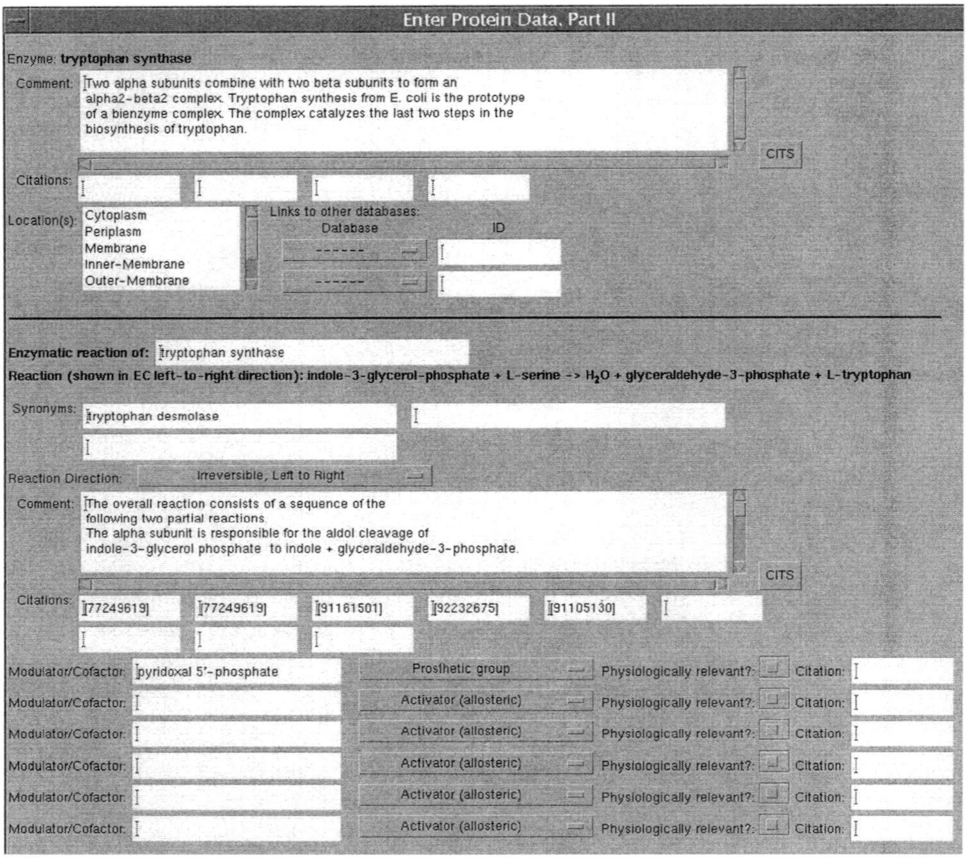

Enter Protein Data, Part II

Enzyme: **tryptophan synthase**

Comment: Two alpha subunits combine with two beta subunits to form an alpha2-beta2 complex. Tryptophan synthesis from E. coli is the prototype of a bienzyme complex. The complex catalyzes the last two steps in the biosynthesis of tryptophan.

CITS

Citations:

Location(s): Cytoplasm, Periplasm, Membrane, Inner-Membrane, Outer-Membrane

Links to other databases: Database ID

Enzymatic reaction of: tryptophan synthase

Reaction (shown in EC left-to-right direction): indole-3-glycerol-phosphate + L-serine -> H_2O + glyceraldehyde-3-phosphate + L-tryptophan

Synonyms: tryptophan desmolase

Reaction Direction: Irreversible, Left to Right

Comment: The overall reaction consists of a sequence of the following two partial reactions. The alpha subunit is responsible for the aldol cleavage of indole-3-glycerol phosphate to indole + glyceraldehyde-3-phosphate.

CITS

Citations: [77249619] [77249619] [91161501] [92232675] [91105130]

Modulator/Cofactor: pyridoxal 5'-phosphate | Prosthetic group | Physiologically relevant?: | Citation:

Modulator/Cofactor: | Activator (allosteric) | Physiologically relevant?: | Citation:

Modulator/Cofactor: | Activator (allosteric) | Physiologically relevant?: | Citation:

Modulator/Cofactor: | Activator (allosteric) | Physiologically relevant?: | Citation:

Modulator/Cofactor: | Activator (allosteric) | Physiologically relevant?: | Citation:

Modulator/Cofactor: | Activator (allosteric) | Physiologically relevant?: | Citation:

FIGURE 3 The Pathway Tools enzyme editor.

The Ocelot data model structures all data as *frames*, which are of two types: classes and instances. Class frames describe general classes of entities, such as the class of all genes, or the class of all pathways. Instance frames describe specific biological entities, such as a specific gene or pathway. The object data model allows the complexities of biological data to be modeled far more compactly than does the relational data model.

Ocelot databases (such as PGDBs) can be stored persistently in Oracle databases and in disk files. The disk-file approach is advantageous for an organization that does not wish to face the complexity of installing and managing Oracle.

However, Oracle storage is required when multiple users want to be able to update a PGDB simultaneously. The Oracle approach also allows faster incremental saving of PGDB updates, in contrast to the disk-file approach, which requires that a PGDB be saved in its entirety when it has undergone any change. The performance of Ocelot has thus far been adequate for managing 10 PGDBs for microbial genomes simultaneously on low-end Sun workstations such as the Ultra-5.

DISCUSSION

The Pathway Tools will be particularly useful to organizations that are interested in metabolic enzymes as a class of drug targets. Metabolic pathways provide the energy and manufacturing plant of the cell. Compounds that interfere with those processes will disrupt cell growth.

We advocate a two-phased approach to microbial drug design. Phase I is the search for *essential in vivo metabolic pathways:* pathways whose function is essential for microbial growth in the host. Phase II is the search for targets within essential *in vivo* pathways.

This approach is advantageous for two reasons. First, by dividing the search for targets into two phases, we drastically reduce the size of the target search space. A strategy that simply considers each gene product as a potential target faces a target space of thousands of genes. In contrast, the two-phased pathway approach reduces the initial search to consider hundreds of essential pathways. If we assume that 10–20 essential pathways are identified, and that each pathway contains 10 gene products, Phase II will consider 100–200 proteins as targets. The second advantage is that knowledge of pathway topology is extremely valuable in Phase II. For example, targets that occur in multiple essential pathways are preferred over targets in a single essential pathway. Targets without multiple isozymes are preferred over those with multiple isozymes. Targets that occur in a location within the pathway that cannot be circumvented by another branch of the pathway are preferred over targets in parallel branches. Targets that occupy pathway holes (pathway steps whose enzymes have not yet been determined) are preferred over identified enzymes because the enzyme that fills a pathway hole (if it can be found within the genome) is unlikely to have been patented.

As well as providing global insights about the biochemistry of the organism, the pathway-prediction process is a method for validating the genome annotation produced through sequence analysis. Pathway holes indicate what gene functions still remain to be identified within the genome, whereas singleton enzymes in pathways indicate possible false-positive function predictions, since it is unlikely that an organism would contain a single enzyme in the middle of a pathway.

The metabolic Overview diagram can be used to analyze gene-expression and protein-expression data in a pathway context. For example, by using this tool to analyze gene-expression data from growth of a microorganism in the presence of a lead compound, the Overview can be used to detect compensating pathways that might become active under those conditions. The Overview can also be used for comparative pathway analysis. A menu-driven query interface allows the user to visualize reactions that are common to one specified set of organisms, but absent from a second set, in order to identify targets present in a preferred group of organisms.

SUMMARY

Pathway Tools is a powerful software package for operating on Pathway/Genome Databases that provides a pathway framework for antimicrobial drug discovery. The capabilities of the software include prediction of the metabolic network of an organism in the form of a PGDB, publishing of the PGDB on a Web site, comparative analysis operations and visualization of expression data in a pathway context, and interactive curation of PGDBs.

ACKNOWLEDGEMENTS

This work was supported by Grant 1-R01-RR07861-01 from the National Center for Research Resources.

REFERENCES

1. P Karp et al., *Nucleic Acids Res, 28(1)*:56, 2000.
2. P Karp, In: *Nucleic Acid and Protein Databases and How to Use Them*, London: Academic Press, 1999, pp. 269–280.
3. P Karp, *Comp. Funct. Genom.* 2:25, 2001.
4. S Brenner, *Trends Genet. 15(4)*:132, 1999.
5. C Ouzounis, P Karp, *Genome Res. 10*:568, 2000.
6. P Karp, VK Chaudhri, SM Paley, *J. Intell. Inf. Syst. 13*:155, 1999.

5

Genomic Strategies in Antibacterial Drug Discovery

Christopher P. Gray and Wolfgang Keck
Morphochem AG, Basel, Switzerland

INTRODUCTION

As has been emphasized since the mid-1990s, there is a pressing need for new classes of antimicrobial compounds to circumnavigate resistance problems. The lifetime of second-generation drugs in the clinic is increasingly shorter. One proposal is to seek novel targets for which it is to be assumed that the inhibitors will also be novel. Alternatively, antibiotics that have never been used in the clinic either because of toxicity problems or lack of information as to their mode of action could be reinvestigated using the more advanced tools now at hand.

The quest for new targets has been indirectly facilitated by the human genome project. The development of rapid sequencing techniques led to the complete genomic sequences of free living organisms, first of the *Haemophilus influenzae* genome [1] and then to that of many more bacteria (available at: http://www.tigr.org/tdb/ and http://www.sanger.ac.uk/Projects/Microbes/). This has enabled a comparison of genomes that has in turn allowed the identification of conserved genes. In a first analysis, these conserved genes are considered to be essential for the growth and/or spread of bacteria in general. These analyses not only compare the sequence conservation of the genes but also the phylogenetic conservation, including gene order and the presumed habitat of the organisms. Such comparisons, now termed comparative genomics, have also addressed the metabolic abilities of bacteria sometimes even predicting the total metabolic capacity. One of the first comparative analyses [2] predicted that *H. influenzae* is best suited for anaerobic growth, whereas *in vitro* the bacteria grow faster and to higher densities under aerobic conditions. This difference between the *in silico* analysis and real life is presumably due to the incompleteness of our knowledge as to the nonannotated genes and the regulation of expression. With no further information as to when and where genes are expressed these predictions can only remain hypotheses. That said, it has, however, been possible to identify genes that fulfill specific requirements for individual pathogens, thereby defining new virulence genes.

With the many genome sequences available, comparative genomics is also addressing the problems of the functional annotation for gene products encoded by novel coding sequences (CDSs) using the information gleaned from the molecular architecture of the genes and domain structure revealed by X-ray crystallography. Functional annotation is not only a problem for the gene products encoded by novel CDSs but is also required for many earlier database submissions. Either the entry only reflected the researchers' direct interests or had no specific meaning as to function. An example of the latter can be found in an *Escherichia coli* protein annotated in the SwissProt database as "a histone like protein" (*hlpA*), "an outer membrane protein" (*ompH*) or "a seventeen kilodalton protein" (*skp*). Only the first of these annotations provides any information as to a possible function. The last annotation not only provides no information as to function but also could, in fact, be misleading as the *H. influenzae* homolog is ~21 kDa.

Generally annotation of the genome sequence databases was performed only at the time of publication. Apart from the genomes of bacteria that are considered to be of commercial value the newer findings for gene products of several bacteria can be accessed in the curated public database (available at: http://www.pasteur.fr/externe). For *E. coli* several websites (e.g., http://web.bham.ac.uk/bcm4ght6/genome.html) are regularly updated but for other bacteria the data must be sifted from the literature.

As yet, the information gained from the genomes has not, to our knowledge, initiated any programs for drug discovery directed at gene products that were previously unknown, i.e., "conserved hypothetical proteins." The rationale for this is clear; the development of an assay for a protein for which the function is unknown is impracticable. The genomics initiative has, however, been instrumental in the validation of enzymes and biosynthetic pathways that have until now been largely ignored. An example of this is fatty acid biosynthesis. Although inhibitors such as thiolactomycin, cerulenin, triclosan, and diazaborines have been known for many years, it is only recently that the targets for these antibiotics have been identified. The identification of the genes encoding the enzymes required for fatty acid biosynthesis in the different bacteria has led to an understanding of the antibacterial activity of these inhibitors and also to the validation of the pathway as a target. Another area where genomics is making an impact on target selection is in the clustering of targets according to their potential catalytic mechanisms or substrate similarity. These properties can be inferred by examining the domain homologies between all the CDSs of a genome.

The science of comparative analysis is relatively young but programs are continuously being developed which, together with the technologies for analyzing expression, will accelerate the process. It is only a matter of time before the functions of nonannotated gene products, identified as essential in pathogenesis, are elucidated and take their place as targets for antimicrobial research.

The development of antibiotics has passed through several phases. After Fleming's discovery of penicillin microorganisms were screened in whole-cell assays for the production of active agents. If found, all that was required was to determine the antibacterial spectrum, stability, toxicity, and the best formulation. The structure of such natural products was, with time, elucidated, which sometimes allowed the chemist to modify the product to enhance some desired property. In most cases, even where possible, this modification has proven to be difficult due to the structural complexity of the starting compound.

The screening of natural products is a slow process often resulting in the "rediscovery" of the same product from phylogenetically different organisms. Other problems associated with screening natural products from micro-organisms are the culture conditions for the biosynthesis of the desired products and the limited proportion of available organisms that can be cultured. There is still a bias to grow cultures containing a single strain, a situation seldom found in nature and most certainly a major factor in our lack of success in realizing the full potential of natural products.

Since the observation that the dyestuff Prontosil rubrum is metabolized in the liver to produce the antibiotic sulfanilamide, natural product screening has been, at first, supplemented with and then largely superseded by the screening of chemical libraries. With more information as to which processes are essential for bacteria the whole-cell assay has been replaced with the enzyme assay, the emphasis always moving toward a better understanding of the structure of the enzyme/ligand complex. Although there have been successes, the screening of large chemical libraries cannot be considered as a cost-effective exercise, even with the trend to high-throughput screening. Very few "hits" are identified in these libraries mainly because they are the result of directed synthesis for defined projects and lack diversity. Paradoxically this lack of diversity has led to a renaissance in attempts to screen natural products.

Combinatorial chemistry, producing millions of new compounds, was also introduced to overcome the lack of diversity in standard libraries but was soon found to be impracticable in that the resynthesis and isolation of "hits" did not result in the desired activity. In many cases it was suspected that the activity was the result of undefined impurities that could not be reproduced in the resynthesis.

More recently there have been attempts to rationally design inhibitors. These attempts have met with only partial success because, although the inhibitor may be improved as to its *in vitro* activity, the modifications required are not always compatible with properties such as solubility, lack of serum absorption, and so on. Using a combination of computer-selected "needles" and multicomponent reactions (MCR) we are now able to combine rational design and diversity. Multicomponent reaction is the name given to a form of combinatorial chemistry first described in 1838 in which a mixture of diverse derivatives of two to five precursor classes will react sequentially to give defined classes of end products [3]. Starting with the crystal structure of the target or, where possible, the target/ligand complex, a computer program that has been developed in-house suggests lead compounds by virtually docking compounds from chemical databases. In a second step the program proposes a series of reactions that can be carried out through MCR to build diversity around the lead compounds. This allows the introduction of diversity at any position in the lead molecule. Using the appropriate assays to analyze the efficiency of the desired property and X-ray crystallography to determine the structure of the target/product complexes, the products are assessed and the results fed back into the program that then refines the next set of MCR. This process of introducing diversity and feedback loops accelerates the development of a lead molecule to a drug candidate for clinical evaluation.

TARGET IDENTIFICATION

The identification of the target molecule of an antibiotic is nowadays a prerequisite for the development of a drug. Where possible the chemist and in particular the medicinal chemist try to design their compounds with reference to the structure of the target or, better still, the target/ligand complex.

The determination of the mode of action of an antibiotic is also of interest as an understanding of the interaction of the drug and its molecular target could generate new hypotheses as to how to improve or synthesise entirely novel inhibitors. It should be further noted that any well-characterized inhibitor of a cellular process is a useful tool to enhance our knowledge of biochemistry that in the long term will lead to new unexplored targets for antimicrobials.

Over the years various methods of studying inhibitors and inhibitor/cell interactions have been used to identify or verify targets at the molecular level. For simple structures the chemical class of an inhibitor can act as a preliminary indication as to its target or target pathway. Analogies of the structure to metabolic intermediates, cofactors, nutrients, and so on, may quickly lead to hypotheses as to possible classes of target but mostly it is only in retrospect that the significance of the analogy is appreciated. In the case of puromycin, the similarity to the aminoacyl terminus of tRNA was a major determinant in understanding the action of this antibiotic. Another example whereby analogy contributed to the identification of the target pathway was when the similarity of the sulfonamides to *p*-aminobenzoic acid, a metabolite in the biosynthesis of folic acid, was noticed.

With the elucidation of the folate biosynthetic pathway, dihydropteroate synthase could be recognized as the target enzyme, although even today we do not fully understand the mechanism of inhibition.

Observing the physiological state of cells exposed to the inhibitor also provides information as to the mode of action. Clearly lysis indicates a perturbation of the cell wall directing attention to cell-wall constituents and their biosynthetic pathways. Cessation of growth can be analyzed as to whether it is reversible upon removal of the antibiotic or the addition of various supplements (bacteriostatic vs. bactericidal).

Biochemical analysis of cellular constituents such as nucleotide levels, the kinetics of nutrient uptake, or the kinetics of macromolecular biosynthesis is also useful in generally discerning the possible pathways that could be affected. At a more advanced stage, *in vitro* assays probing suspected biochemical systems can be analyzed. However, these may prove negative when metabolism of the drug is necessary for the interaction or when the target is part of a complex that is disrupted upon purification.

With the advent of total expression analysis we have the possibility of rapidly examining the changes effected upon the cell by an inhibitor. Facing a variety of growth conditions (e.g., starvation, heat, anaerobiosis, and toxins), cells have developed highly elaborate regulatory networks to adapt to different environments, temperatures, nutrients, and so on. Many of these adaptive responses take place at the level of gene expression. Thus, the gene expression pattern reflects the underlying growth condition. Inhibition of a particular function with an antibiotic is comparable to exposure to a toxin and the cell's response will also be reflected in the gene expression pattern. The response often has more than one component but will always include a specific response which results from the cell's direct attempt to overcome the inhibition. Although the mode of action of many antibacterials has been described in some detail the study of their effects on the delicate networks of metabolism, gene regulation, cell cycle, and so on, is still in its infancy. However, with experience, an interpretation of these changes will lead to a rapid determination of the target pathway and, with further experimentation, validation of the target molecule itself. Two approaches relying on the analysis of the response of bacteria after exposure to the inhibitor are presently being employed. Either the change in the expression pattern in response to the inhibitor is compared to a database of responses or an hypothesis is proposed based on the changes observed in the gene expression/protein synthesis/metabolites involved in particular cellular processes. The first approach is the easiest but requires a response pattern for an antibiotic that is disrupting the same cellular process. Even when such a response pattern does exist, the new compound does not necessarily elicit a response pattern that can be identified as such. This also applies to compounds derived from chemical programs directed at a defined target. Difficulties that arise in this comparison approach involve the concentration level of inhibitor and the time point after addition of the inhibitor at which the cells should be analyzed. We have tried to standardize this by using a concentration of inhibitor that reduces the growth rate by a factor of 2 and to perform a kinetic study. This approach was reasonably successful for analyzing the proteome but often resulted in changes that were too subtle when analyzing expression patterns. In an experiment performed as above, comparing the responses of three newly developed dipyrimidines with good *in vitro* activity against dihydrofolate reductase to the response pattern of trimethoprim (TMP), only one of the new compounds resulted in a response pattern that had any similarities to that of TMP. The responses elicited by the other two inhibitors cannot be interpreted at this stage and would require analysis at many concentrations and times after addition of the inhibitors. We have also noted that many of the classical antibiotics for which the target has been described produce phenotypes at the mRNA and protein level which indicate that they, in fact, inhibit more than one cellular process. It is a philosophical argument as to whether this serendipitous situation is advantageous. The major consideration with such findings, using new compounds, is the potential that these secondary affects will prove to be toxic to the eukaryotic cell, thereby eliminating there usefulness at a later and therefore more expensive stage of the drug discovery process.

Although, as already mentioned, there are no drug discovery programs involving gene products derived from CDSs of unknown function, there are still good reasons to identify new targets. Resistance development is not the only reason that antibiotics sometimes fail to clear infections. Targets are usually described as being essential proteins but the term *essential* must be qualified to "essential for a particular physiological state of the cell." Bacteria can and do change their complete metabolic processes to suit their environment. This can lead to the cell entering a phase where it does not actively grow, thereby rendering genes normally essential for growth nonessential. In such a state antibiotics that inhibit cell wall biosynthesis (e.g., the penicillins), DNA replication (e.g., the quinolones), or even protein synthesis (e.g., many of the macrolides) become ineffective. Infections that do not respond to antibiotic treatment are most commonly found for bacteria that live intracellularly (e.g., *Chlamydia sp.*) or form biofilms (e.g., *Staphylococcus aureus*) but can also occur when the infectious agent grows planktonically (e.g., *Streptococcus pneumoniae*). At present we know very little about how and why the cells enter this state and what is necessary to maintain it.

The best studied systems, because of their amenability to experimentation, are biofilms. It would appear that bacteria in most ecosystems prefer to live in biofilms associated with surfaces [4]. Biofilms cause chronic infections accounting for approximately 65% of human infections, for example, lung infections in cystic fibrosis patients, prostate infections, and endocarditis. Microbes in biofilms exist in layers that adhere to a surface and are protected from the cells of the immune system. The resistance of bacteria within biofilms to antibiotics would not, as previously thought, appear to be a problem of penetration but rather, as mentioned above, a reprogramming of gene expression [5]. Experimentally it was shown that bacteria in the center of the biofilm are less susceptible than bacteria growing on the outside because they enter an anaerobic state, which spurs them to downshift and stop growing. Access to the genomic sequence of *E. coli* has allowed the identification of many genes in a study using random insertion mutagenesis, with a promoterless *lacZ* Mu phage construct, to assess changes in gene expression in biofilm cells versus planktonic cells [6]. Again using the genomic sequence, in this case, from *S. aureus*, microrepresentational-difference analysis (micro-RDA) could be employed to identify genes that are typically expressed in biofilms [7]. The results from these studies, plus many others that do not address the whole genome, all point to the cells responding to a signaling process known as quorum sensing [8]. Due to species specificity, quorum sensing itself as a target process will most likely only lead to narrow-spectrum antibiotics. The analysis of the expression patterns of the cells in biofilms from many different species may, however, result in antibiotic targets with broader spectra.

At the time of this writing, there have been few attempts to apply genomic technologies to the analysis of the expression in bacteria in which the persistence is either intracellular or planktonic. Both states require *in vivo* models and this limits which technologies are practical. One of the few publications that does attempt a more "global" approach [9] used signature tagging, identifying a gene in *Brucella abortus* that is homologous to a gene already known from *Mycobacterium tuberculosis* to be essential for persistence. This suggests that some mechanisms for long-term persistence may be shared among chronic intracellular pathogens.

TRANSCRIPTOME OR PROTEOME: DIFFERENT VIEWS OF COMPLEXITY

Using the technologies that have evolved with the genome initiatives it is now possible to quantify all the transcripts or hundreds to thousands of the proteins present in the cell at any particular time point. The question arises as to which technologies are appropriate for the analysis of bacterial responses. Although the total phenotype of a cell is determined by the proteome it is easier to analyze the transcriptome. However, it must be realized that due to posttranslational modifications, changes at the protein level are not necessarily reflected by changes at the RNA level. Another consideration when analyzing the responses of the cell via the expression pattern is the half-life of the different mRNAs that will be dependent on the individual sequences together with the RNA degrading complexes present at any particular time. The regulation of the type and number of these degrading complexes must also be considered as a target of what we term "the secondary effects" of an antibiotic.

At present, it is only possible to quantify the amount of a particular mRNA as a fraction of the total mRNAs. If a particular mRNA in the induced state is degraded as fast as it is transcribed and translated, then the level of this message will not change as a fraction of the total. This scenario is of course an extreme situation but is mentioned to illustrate that the level of mRNA measured is not necessarily reflecting the level of transcription. Concrete examples of this phenomenon have been encountered in the differences between measuring proteins as a fraction of the total compared to what is actually being translated within a short time frame. This is demonstrated in Figure 1, which is taken from a series of experiments studying the responses of *H. influenzae* to TMP. As can be seen the changes in the levels of expression measured as a fraction of the total protein present in the cells are not as dynamic as when the *de novo* changes are measured. This figure also illustrates the additional information that can be obtained from studying the proteome in that the two forms of the protein, presumably the result of posttranslational modifications, do not always change to the same extent or even in the same direction. In these five examples only 2 spots could be identified as representing the proteins; however, some proteins were represented by at least

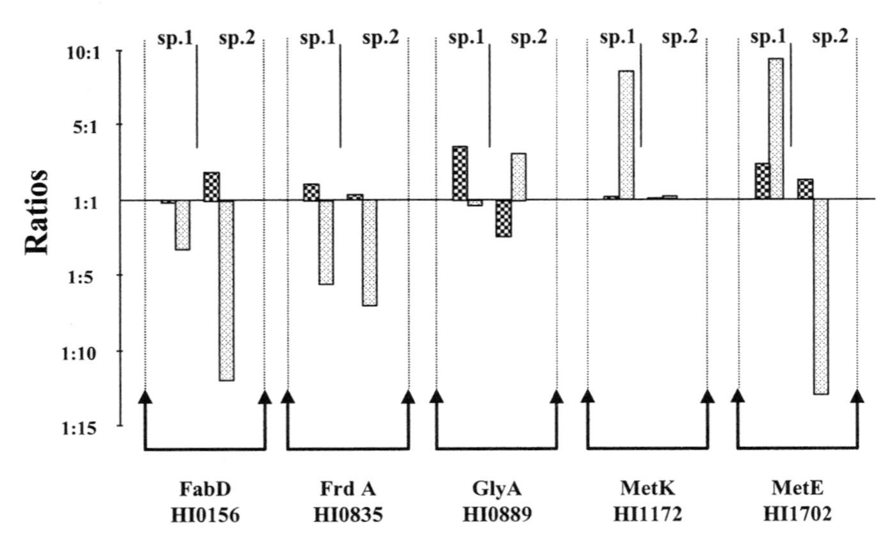

Figure 1 Dynamics of expression. The ratios for the changes in the intensity of five different *H. influenzae* proteins after the cells had been treated for 30 min with trimethoprim are shown. For each protein, two spots (sp. 1 and sp. 2) could be identified in the 2D-PAG image (see text). Cells were labeled either continuously (▩) or during 3-min pulsing (▤) with [^{35}S]methionine.

5 spots. The phenomenon of a gene product being represented by more than one spot in α two-dimensional polyacrylamide gel (2D-PAG) is often encountered and in the case of eukaryotes it has been demonstrated that many gene products are represented by up to 10 spots [10]. The characterization of the various forms of the proteins represented by the different spots will add yet another layer of complexity to the analysis of expression.

Many methodologies have been developed to analyze the changes in the levels of mRNAs induced by external stimuli. Each method has its perceived advantages and disadvantages but very few groups have published direct comparisons between more than two methodologies. It has, however, been noted that there are differences in the relative values obtained using different methodologies for the detection of the RNA. We have experienced this with values observed for several genes analyzed by either the AffyMetrix Chip or by Northern methodologies. In the best-studied example Northern analysis resulted in values that were comparable to those determined from the proteomics data. It would be preferable to be able to measure the synthesis and degradation of both the mRNAs and the proteins, but as mentioned for the former this is not yet possible and for the latter we can only measure a subset.

For many years a major distinction cited between eukaryotic and prokaryotic mRNA was the polyadenylation of the former. We now know that prokaryotic RNAs can also be polyadenylated but to a much lesser extent and that this polyadenylation is a signal for degradation. Whether all, or only subsets, of RNAs are flagged in this manner is, at present, unknown. However, an analysis of these polyadenylated RNAs could open a window into the dynamics of RNA turnover.

MODEL ORGANISMS

The two most extensively studied bacterial species are undoubtedly *E. coli*, as a gram-negative, and *Bacillus subtilis*, as a gram-positive. The genomic sequences of these two organisms were, however, unavailable when we initiated our microbial genomics program. The first free-living organism for which the entire genetic information became available was *Haemophilus influenzae* [1], which is a small, nonmotile, gram-negative bacterium. Its only natural host is the human, where it is found as part of the normal commensal flora of respiratory and genitourinary tracts [11]. Infection is caused by invasion of the blood stream and spread from the respiratory tract, with strains of capsular type B being the most invasive. Although the introduction of vaccines based on type B capsular antigen has greatly diminished the incidence of pediatric otitis media [11], *H. influenzae* infections are still relatively common in children. Its importance as a pathogen, its relatively high susceptibility to antibiotics, and its amenability to genetic manipulation make *H. influenzae* an attractive model organism.

As a representative of Gram-positive bacteria, we chose another important human pathogen, *Streptococcus pneumonia*, which causes invasive infections such as sepsis, meningitis, and pneumonia [12]. As with *H. influenza*, *S. pneumonia* has a relatively small genome and is susceptible to antibiotics and amenable to genetic manipulation. In a collaboration between Hoffmann-La Roche and Human Genome Sciences/TIGR, approximately 95% of the genomic sequence of the strain R6, a laboratory strain, was determined and posted on the TIGR database in 1997 to become one of the first publicly available Gram-positive genomes.

RESPONSE PROFILING IN PRACTICE

We have used *H. influenzae* in a number of studies examining protein synthesis, expression or both protein synthesis and expression in response to different antibiotics.

In one series of experiments [13] we used 2D-PAGE alone to analyze the effects of a transcriptional inhibitor (rifampicin) and several translational inhibitors (chloramphenicol, erythromycin, fusidic acid, puromycin, and tetracycline), including representatives of the aminoglycosides (kanamycin and streptomycin). Streptomycin was also chosen because of its particular characteristics with respect to mode of action and resistance development [14]. Data extracted from the literature proposes the following mechanisms of inhibition: Chloramphenicol prevents the peptidyltransferase reaction [15], erythromycin is believed to act by blocking the translocation of peptidyl-tRNA from the A to the P site [16], fusidic acid inhibits the same step by binding to elongation factor G [17], and tetracycline prevents binding of aminoacyl-tRNA mainly to the A site [18]. Finally, puromycin is an aminoacyl-adenosine analog and leads to premature chain termination [16]. A protein map of *H. influenzae*, accounting for more than 500 unique proteins [19,20], facilitated this study.

The responses to the arrest of protein synthesis, either through inhibition of different stages of the protein synthetic process itself or through inhibition of the synthesis of mRNA, were compared to evaluate whether the response patterns are indicative of the different modes of inhibition. It could be shown that the induction of the synthesis of components of the transcriptional and translational machinery is a characteristic common to all transcriptional and translational inhibitors included in the study with the exception of the aminoglycosides. The relative rate of synthesis of these proteins, therefore, potentially provides a diagnostic tool for the inhibition of transcription or translation. The failure to detect this response in cells treated with aminoglycosides is most likely due to their action against multiple targets [14]. However, using a simple comparison program, it was possible to show that responses to aminoglycosides were more similar to the transcriptional/translational inhibitors than to antibiotics acting upon targets involved in different cellular processes.

In a second study, responses to inhibitors of DNA gyrase were examined using both the proteomic and transcriptomic approaches. The goals of this study were threefold. As well as investigating the responses of the two classes of antibiotics, we wished to cross-validate the two technologies, transcriptomics and proteomics, evaluating to what extent the combined use of the two technologies could enhance the power of expression analysis.

DNA gyrase (E.C. 5.99.1.3.), a prokaryotic topoisomerase II enzyme essential for viability, consists of two subunits, A and B, the active enzyme being an A2B2 tetrameric complex (for review see 21). The enzyme has no direct mammalian counterpart and is the only enzyme known to be able to introduce negative supercoils into DNA by using the energy derived from ATP hydrolysis. A key step in this supercoiling reaction is the DNA gyrase-mediated cleavage of DNA. It has been shown that a class of inhibitors binding to subunit A, the quinolones and the pyrimido[1,6-α]benzimidazoles, interrupt the cleavage and resealing cycle at the cleavage step [22,23]. This is effected through the formation of a stable ternary complex consisting of the enzyme, DNA, and the inhibitor and results in DNA damage, which in turn blocks replication and transcription (for review see 24). As a consequence, the expression of DNA repair systems, mainly the SOS system, is induced [25]. It has also been noted that, depending on the bacterial species and on the quinolone, preferential inhibition of topoisomerase IV can be observed [26,27,28]. *In vivo*, quinolones also have additional effects that go beyond the inhibition of topoisomerases.

A second class of DNA gyrase inhibitors, the cyclothialidines and the coumarins, bind to the ATP binding site located in the subunit B, thereby inhibiting the supercoiling activity of the enzyme but leaving the DNA otherwise intact [29].

Although these inhibitors are not *per se* bactericidal, by inhibiting the activity of DNA gyrase they indirectly influence supercoiling. The initiation of transcription of many genes is sensitive to DNA supercoiling, often exhibiting an optimum with respect to the degree of supercoiling [30]. As expected, this difference in the mode of action was reflected in the expression patterns resulting from the cells' responses to the two inhibitors.

At low concentrations of the quinolone ciprofloxacin, only a few genes were altered in their level of expression. Most of these genes are annotated as belonging to DNA repair processes and include genes involved in SOS repair (*recA*, *uvrA*, and *lexA*) as expected. At higher concentrations of ciprofloxacin, the level of expression changed for many genes, most probably indicating secondary affects. It was of interest that the expression level of the gene *parC* was not drastically changed which would indicate that topoisomerase IV in *H. influenzae* is not a target for this quinolone. This is in contradiction to the findings that mutations in *parC* confer resistance to ciprofloxacin [31,32].

At low concentrations of the coumarin novobiocin (1X MIC), only a few genes were altered in respect to their level of expression. Most significantly, these included DNA gyrase (subunit B) and topoisomerase I, a result that can be interpreted as the cells' attempts to compensate for the enzyme inhibition and to maintain optimal supercoiling. At higher concentrations (10X MIC), approximately 37% of the genome showed, at least at one time point, increased or decreased expression rates. These presumably secondary affects were considered to be nonspecific inhibitory activity of ATPases that have similar binding sites to the gyrase.

Not surprisingly, the sensitivity and reproducibility of this analysis was better using the oligonucleotide chip technology than that obtained using proteomics. The changes in the values measured for transcripts were within three orders of magnitude, whereas those measured for proteins varied within five orders of magnitude. However, although the changes were qualitatively similar, there were quantitative differences in the responses detected by protein quantification compared to those detected by mRNA quantification. This highlights the importance of combining both technologies to obtain information as to the level (transcriptional or translational) at which the regulatory mechanisms act. Moreover, posttranslational modifications constitute an important additional level of regulation and can only be studied by proteome investigations. The detection of proteins present as multiple spots underscores this point (see above).

Finally it could be concluded from this set of experiments that, as expected, lower concentrations of inhibitor result in a more specific response. The concentration of inhibitor used to achieve specificity will depend on what levels of change in the specific signal can be measured compared to the changes in the nonspecific signals. An example has already been mentioned above for inhibitors of DHFR. In that case, even at what were considered to be low concentrations, the changes in the specific signals were minor compared to those in the secondary/nonspecific signals. Clearly, a wide range of concentrations will have to be analyzed for every inhibitor.

Whereas creating a database of responses to antibiotics that are commercially available can be carried out using large culture volumes, miniaturization is required for analyzing the compounds produced by MCR chemistry. Miniaturization of the cultures obviously results in a miniaturization of the amount of RNA that can be recovered and the labeling process therefore requires an amplification step. The major concern with the amplification of cDNA is related to PCR. There is a real risk of biasing the results by the specific amplification of fragments. Recently methods have been developed in the analysis of tumor samples that use a combination of PCR and *in vitro* transcription for the specific amplification of mRNA [33]. It was demonstrated that, using such combinations, it is possible to amplify without introducing a bias. We are applying an adaptation of this methodology to amplify labeled cRNA from bacterial total RNA.

ALTERNATIVE OR COMPLEMENTARY TECHNOLOGIES TO TRANSCRIPTOMICS

Resistance to an antibiotic can result from the overexpression of the target gene. A relatively common example of this form of resistance, observed in clinical isolates, is the overexpression of the DHFR gene in response to treatment with trimethoprim. Either the expression is upregulated by mutations of the promoter region [34] or the gene is located on a multicopy plasmid [35]. This principle has been exploited to isolate the DHFR gene from various organisms [36]. A second example whereby the principle of overexpression has been utilized to isolate the target gene of an antibiotic is for thiolactomycin [37].

In experiments to investigate the usefulness of this principle for isolating the resistance genes (which could include the target gene itself) a library of *E. coli* fragments cloned into pUC were plated out on agar containing well-characterized antibiotics at concentrations slightly higher than their relevant MIC. The plasmid inserts from clones able to grow on these plates were sequenced and in four cases the target gene was indeed included among the sequences. An analysis of the sequences, however, demonstrated the limitations of this approach in that the target gene could be represented in only a small fraction on the clones or sometimes not at all. This can easily be explained for antibiotics for which the target is a protein complex involving genes with different loci as for example in the case of quinolones and DNA gyrase but the explanation is not always so clear. This clonal approach involves the sequencing of many fragments to obtain a statistically reliable result and does not take the fitness of the bacteria into account. We are adapting this methodology to culture, analyzing inserts that convey an advantage to the cell using *E. coli* oligonucleotide arrays.

The ability of cells to survive in sub MIC levels of antibiotics is dependent not only on the expression of the target gene or resistance mechanism but also on the changes in the expression of other genes. Some of these genes, although nonessential under normal growth conditions, become essential when the cell is challenged with the antibiotic. In order to identify these genes we are analyzing libraries in which genes have been disrupted by random transposition. The disappearance of cells containing a disruption in a particular gene indicates the requirement for that gene for growth in the antibiotic. Knowledge of these genes also provides pointers as to the mechanism of the antibiotic.

CONCLUDING REMARKS

For many years, few investigators considered the prokaryotic cell as more than a convenient sack containing genes and gene products. This perception was reflected in numerous studies where genes were investigated as individual entities. Various approaches using cloning, chemical/UV-induced mutation, transposition, or antibiotics all sought a clean phenotype that was then considered to be the result of the change in the expression of a single gene. This philosophy is particularly entrenched in the textbook explanations of antibacterial activity. A more detailed study, however, often reveals secondary targets that are important for the efficacy of a particular drug. Using the technologies that have been developed since the late 1990s we can now put genes back into context and examine the true complexity of cellular processes and the disruption of such.

As with the genes, bacterial species have, until now, been mostly studied in isolation and under optimal growth conditions. A single species is seldom the case in many indications, where even an out-growing pathogen is but one of many species at the site of infection [a recognized problem for bacterial diagnostics [38]. Bacteria are opportunistic, utilizing any nutrients or beneficial factors, including genetic information [39], in the environment in which they find themselves. Their environment also determines their physiological state, which is unlikely to be comparable to that achieved by growth *in vitro*. The interactions between species and their relative physiological states are areas of research that can now be explored, perhaps leading to antibacterial research focused more toward indications rather than species.

REFERENCES

1. RD Fleischmann, MD Adams, O White, RA Clayton, EF Kirkness, AR Kerlavage, CJ Bult, J-F Tomb, BA Dougherty, JM Merrick, K McKenney, G Sutton, W FitzHugh, C Fields, JD Gocayne, J Scott, R Shirley, L-I Liu, A Glodek, JM Kelley, JF Weidman, CA Phillips, T Spriggs, E Hedblom, MD Cotton, TR Utterback, MC Hanna, DT Nguyen, DM Saudek, RC Brandon, LD Fine, JL Fritchman, JL Fuhrmann, NSM Geoghagen, CL Gnehm, LA McDonald, KV Small, CM Fraser, HO Smith, JC Venter. *Science 269*:496–512, 1995.
2. RL Tatusov, AR Mushegian, P Bork, NP Brown, WS Hayes, M Borodovsky, KE Rudd, EV Koonin. *Curr. Biol. 6(3)*:279–291, 1996.
3. A Dömling, I Ugi. *Angew. Chem. 112*:3300–3344, 2000.
4. JW Costerton, PS Stewart, EP Greenberg. *Science 284*:1318–1322, 1999.
5. JN Anderl, MJ Franklin, PS Stewart. *Antimicrob. Agents Chemother. 44(7)*: 1818–1824, 2000.
6. C Prigent-Combaret, O Vidal, C Dorel, P Lejeune. *J. Bacteriol. 181(19)*:5993–6002, 1999.
7. P Becker, W Hufnagle, G Peters, M Herrmann. *Appl. Environ. Microbiol. 67(7)*: 2958–2965, 2001.
8. TR de Kievit, BH Iglewski. *Infect Immun. 68(9)*:4839–4849, 2000.
9. PC Hong, RM Tsolis, TA Ficht. *Infect Immun. 68(7)*:4102–4107, 2000.
10. M Fountoulakis, JF Juranville, P Berndt, H Langen, L Suter. *Electrophoresis 22(9)*: 1747–1763, 2001.

11. JZ Jordens, MP Slack. *Eur. J. Clin. Microbiol. Infect Dis. 14(11)*:935–948, 1995.
12. EA deVelasco, AF Verheul, J Verhoef, H Snippe. *Microbiol. Rev. 59(4)*:591–603, 1995.
13. S Evers, K Di Padova, M Meyer, H Langen, M Fountoulakis, W Keck, CP Gray. *Proteomics 1*:522–544, 2001.
14. A Dalhoff. *Antibiot. Chemother. 39*:182–204, 1987.
15. D Drainas, DL Kalpaxis, C Coutsogeorgopoulos. *Eur. J. Biochem. 164*:53–58, 1987.
16. E Gale, E Cundliffe, P Reynolds, M Richmond, M Waring. *The Molecular Basis of Antibiotic Action*. London: Wiley, 1981.
17. E Cundliffe. *Biochem. Biophys. Res. Commun. 46*:1794–1801, 1972.
18. U Geigenmuller, KH Nierhaus. *Eur. J. Biochem. 161*:723–726, 1986.
19. M Fountoulakis, H Langen, S Evers, CP Gray, B Takàcs. *Electrophoresis 18*: 1193–1202, 1997.
20. H Langen, B Takacs, S Evers, P Berndt, HW Lahm, B Wipf, CP Gray, M Fountoulakis. *Electrophoresis 21*:411–429, 2000.
21. A Luttinger. *Mol. Microbiol. 15(4)*:601–606, 1995.
22. DC Hooper, JS Wolfson. *Eur. J. Clin. Microbiol. Infect. Dis. 10(4)*:223–231, 1991.
23. C Hubschwerlen, P Pflieger, JL Specklin, K Gubernator, H Gmunder, P Angehrn, I Kompis. *J. Med. Chem. 35(8)*:1385–1392, 1992.
24. A Maxwell. *Biochem. Soc. Trans. 27(2)*:48–53, 1999.
25. RA VanBogelen, PM Kelley, FC Neidhardt. *J. Bacteriol. 169(1)*:26–32, 1987.
26. H Fukuda, K Hiramatsu. *Antimicrob. Agents Chemother. 43(2)*:410–412, 1999.
27. TR Shultz, JW Tapsall, PA White. *Antimicrob. Agents Chemother. 45(3)*:734–738, 2001.
28. VE Anderson, N Osheroff. *Curr. Pharm. Des. 7(5)*:337–353, 2001.
29. A Contreras, A Maxwell. *Mol. Microbiol. 6(12)*:1617–1624, 1992.
30. JC Wang, AS Lynch. *Curr. Opin. Genet. Dev. 3(5)*:764–768, 1993.
31. DJ Biedenbach, RN Jones. *Diagn. Microbiol. Infect Dis. 36(4)*:255–259, 2000.
32. M Georgiou, R Munoz, F Roman, R Canton, R Gomez-Lus, J Campos, AG De La Campa. *Antimicrob. Agents Chemother. 40(7)*:1741–1744, 1996.
33. TR Hughes, M Mao, AR Jones, J Burchard, MJ Marton, KW Shannon, SM Lefkowitz, M Ziman, JM Schelter, MR Meyer, S Kobayashi, C Davis, H Dai, YD He, SB Stephaniants, G Cavet, WL Walker, A West, E Coffey, DD Shoemaker, R Stoughton, AP Blanchard, SH Friend, PS Linsley. *Nat. Biotechnol. 19(4)*:342–347, 2001.
34. FM Sirotnak, RW McCuen. *Genetics 74(4)*:543–556, 1973.
35. D Grey, JM Hamilton-Miller, W Brumfitt. *Chemotherapy 25(3)*:147–156, 1979.
36. AC Chang, JH Nunberg, RJ Kaufman, HA Erlich, RT Schimke, SN Cohen. *Nature 275(5681)*:617–624, 1978.
37. JT Tsay, CO Rock, S Jackowski. *J. Bacteriol. 174(2)*:508–513, 1992.
38. J Garcia-de-Lomas, D Navarro. *Pediatr. Infect. Dis. J. 16(Suppl 3)*:43–48, 1997.
39. JS Kroll, KE Wilks, JL Farrant, PR Langford. *Proc. Natl. Acad. Sci. USA 95(21)*: 12381–12385, 1998.

6

Genomics-Based Approaches to Novel Antimicrobial Target Discovery

Thomas J. Dougherty
Pfizer Global Research and Development, Groton, Connecticut, U.S.A.

John F. Barrett
Merck Research Laboratories, Merck & Co., Rahway, New Jersey, U.S.A.

Michael J. Pucci
Achillion Pharmaceuticals, New Haven, Connecticut, U.S.A.

INTRODUCTION

Before the advent of antibiotics, mortality due to bacterial infections was a major cause of death worldwide [1]. While infectious diseases still play a role in global health issues, the discovery of antibiotics had a revolutionary impact on the practice of medicine, as infectious diseases were now treated on an almost routine basis. The introduction of penicillin was followed by the rapid discovery of several new structural classes of antibiotics in the 1940s and 1950s. Even tuberculosis, the dramatic AIDS-like illness of its day, in which young people succumbed after lingering courses, came under the control of chemotherapy in the late 1940s. These "miracle" compounds, most of which were natural products isolated from soil microbes, were identified by straightforward bacterial growth inhibition screens [2]. Antibiotic resistance was initially a laboratory tool used as a convenient phenotypic marker [3]. Isolated cases of clinical resistance failed to raise major alarms. The discovery of transmissible resistance extrachromosomal elements in the 1950s and 1960s was the beginning of the recognition of serious resistance problems. Today, the resistance problem has grown to encompass virtually all the serious pathogens and classes of antimicrobial compounds [4]. Drug-resistant tuberculosis is becoming a widespread issue of serious concern [5]. Attempts to control resistance by cycling the use of antibiotic classes in the clinic has met with mixed success, due to both prescribing practices and the persistence of resistance genes in the absence of drug selection [5,6]. Almost in parallel to the rise in antibiotic resistance in the 1960s, efforts by the pharmaceutical industry to identify novel classes of antimicrobial agents declined. Only within the past 2 decades has there been the recognition of the increasingly serious nature of resistance [8,9]. The consensus has emerged that as part of an overall strategy to control resistant pathogens, novel antibiotic classes will be an essential part of the solution [10].

Since the 1950s, significant modifications have been made to most classes of antibiotics to increase spectrum and potency and limit their vulnerability to resistance mechanisms. Yet the resistance problems that defeated earlier versions of these compounds are now overtaking the newer members of existing classes. The problems, for example, with increasing quinolone resistance levels and novel types of β-lactamase variants that hydrolyze the newer penicillin and cephalosporin derivatives, are well documented [11–13]. In parallel with efforts to modify existing antibiotics, there was a renewed search for novel antimicrobials. This search has yielded to date only one fundamentally new class of antibiotics, the oxazolidinones, introduced into the clinic in the past 30 years [14]. The formulas for success that unearthed multiple new antibiotic classes in the middle of the last century do not appear to be yielding additional compounds. Clearly, a new approach to discovery is needed. If we accept that, given current trends, drug resistance will continue to erode the utility of existing compound classes, then it becomes clear that innovative strategies are necessary to discover novel antimicrobials and maintain our ability to control serious infections.

The existing structural classes of antibiotics with utility are relatively few and target a small subset of essential bacterial processes [11]. While it can be argued that not every essential function in bacteria is an ideal antimicrobial target, it is clear that identifying the sum total of essential bacterial gene products has the potential to lead to the identification of novel inhibitors of a subset of these reactions. These would in turn expand the number of available antimicrobial classes. It might be argued that if inhibitors of these targets existed, they should have been identified in whole-cell screens in the past, where in essence a microbe's entire set of essential functions are screened. This presumes, however, that such inhibitors are made as natural products in concentrations sufficient to be detected in such screens or that potent inhibitors of one or more of these functions have been synthesized in the laboratory by chemists. With the identification and availability of essential gene products through the use of microbial genomics, it will be possible to establish novel *in vitro* and *in vivo* screens of increased sensitivity to detect inhibitors. It will also be possible to perform detailed structural analyses on purified forms of the targets (protein or otherwise), with the goal of computer-aided design of specific functional inhibitors. These novel targets may also be inhibited by members of the array of new compounds that are generated by combinatorial chemistry. In the course of this review, some of the strategies to identify novel targets for drug development among the newly available microbial genome-based information will be highlighted.

ANTIBIOTIC RESISTANCE

The driving force behind the renewal of interest in antibacterial programs is unquestionably the recognition of the serious and growing bacterial resistance problem. In recent years, there has been a growing appreciation that serious pathogens have acquired a range of resistance mechanisms. For years, the dissemination of R plasmids with multiple resistance determinants was a widespread and well-known phenomenon [15,16]. However, many surprising and unexpected developments in resistance mechanisms were also uncovered. One example is the development of non-β-lactamase forms of penicillin resistance arising from mosaic genes in the penicillin binding proteins (PBPs), the targets of β-lactam antibiotics, that result from exchange of gene segments via transformation in streptococci and *Neisseria* spp. [17]. These novel recombinant genes resulted in variant forms of PBPs, which had reduced affinities for drug and yet retained their essential function in cell-wall synthesis. Along similar lines, methicillin-resistant *Staphylococcus aureus* (MRSA) were found to have acquired an additional PBP that had very low affinity for key β-lactam compounds and could substitute for the native staphylococcal PBPs in wall synthesis [18]. This new PBP, not present in antibiotic sensitive staphylococci, was encoded by part of a larger genetic element only found in MRSA strains. As mentioned previously, the evolution of variant β-lactamase enzymes, the extended spectrum β-lactamases (ESBLs), has been driven by the introduction of newer β-lactam antibiotics specifically designed to be resistant to hydrolysis by known members of this enzyme class [19].

Another unexpected resistance development was the high-level vancomycin resistance gene system in *Enterococci* spp. Prior to this, it was believed that vancomycin, the drug of last resort for certain serious infections, was by its mechanism of action immune to resistance problems. The drug acts by binding to the terminal D-alanyl-D-alanine of the peptidoglycan structural precursor and inhibiting cell-wall assembly. It is clear that this transposon-borne set of resistance genes remodels the bacterial peptidoglycan to remove the vancomycin target of D-alanyl-D-alanine, replacing it with D-alanyl-D-lactate [20,21]. The vancomycin resistance system includes a two-component regulatory system that, in an as yet undetermined way, senses the presence of the antibiotic and turns on expression of the vancomycin resistance genes.

Successful efforts to improve the potency of the fluoroquinolone class has resulted in the selection of increased numbers of clinical strains which have one or more point mutations in the protein subunits of the two targets of this class, DNA gyrase and topoisomerase IV. These mutations are most often at key amino acid residues that apparently interact with the drugs, termed the quinolone resistance determining regions. The mutations reduce the drug affinity for the target protein–DNA complex [12].

Another mechanism that reduces susceptibility to a broad range of antimicrobials is active efflux. This consists of several structurally distinct classes of both narrow and broad specificity efflux pumps. These act to actively pump antibiotics out of cells and thereby reduce the intracellular concentration of antibiotic, resulting in resistance. The discovery of these pumps and their regulation was yet another indication of microbial versatility under stress [22].

Is there a "price" of resistance in terms of reducing the fitness of the microbe expressing resistance? There is evidence that the metabolic expenditure of maintaining a resistance mechanism has a fitness cost in the absence of the antibiotic (e.g., 23), and even a small difference in fitness will lead to loss of the less fit organism from a population over time [24]. Thus, in the absence of drug, a resistant organism should be purged over time by natural selection from the population. This has led to efforts to reduce resistance to an antibiotic class by eliminating or greatly restricting its clinical use for a period of time. However, the idea that removing an antibiotic from the environment will select for a bacterial population that reacquires drug sensitivity has been challenged by the discovery that secondary mutations, either extragenic or intragenic, can restore fitness in resistant organisms [25,26]. The extensive use of antibiotics may have established an environment in which selection not only favors resistant mutants, but also may result in additional mutations that compensate for fitness costs imposed by the resistance rather than drug-sensitive revertants. It has been argued that the rates of fitness compensatory mutations exceed that of true reversions to drug sensitivity, favoring the retention of resistance [27]. This has been demonstrated both in vitro and in mouse virulence models in which antibiotic resistant *Salmonella typhimurium*, which initially were avirulent, reverted to virulence without loss of antibiotic resistance [26]. Although these results may in time be demonstrated to be of considerable importance, it is currently uncertain as to how significant compensatory mutations are to maintaining resistant organisms in a population [27].

Another recent concept in the emergence of resistance is the notion that naturally occurring mutator strains present in microbial populations may accelerate the rates of antibiotic resistance development [28,29]. Mutators are found naturally among bacterial populations and can be the result of deficiencies in mismatch repair (e.g., *mutH*, *mutS*, *mutL*, and *uvrD*), among other mechanisms [30,31]. This can lead to a subpopulation of hypermutable bacteria that could generate antibiotic-resistant mutants at a higher rate than the overall population. In addition, the deficiency in mismatch repair may promote gene exchange via recombination with exogenous DNA, resulting in another source of potential resistance development. Such hypermutable strains could rapidly develop multistep resistance to an antibiotic, as well as resistance to multiple classes of antibiotics. Laboratory evidence exists that indicates that the mutator phenotype, in the absence of selection pressure, leads to loss of multiple gene functions and reduced fitness of the mutator population [32]. There is evidence, however, that hypermutable strains may have an advantage in the rapidly changing environment of the host. A recent finding has been of very high frequencies (36%) of hypermutable strains of *Pseudomonas aeruginosa* in cystic fibrosis patients (CF) [33]. These strains were found to have higher resistance frequencies to a broad range of antibiotics. The CF patient is subjected to multiple antibiotic regimens, and the shifting lung environment may favor organisms that can rapidly adapt to changes. The finding that mutators are also present at high (>1%) incidence in pathogenic *Escherichia coli* and *Salmonella enterica* isolated from food related outbreaks indicates that there may be an advantage to having mutators in the population that permits rapid variation in the shifting environment of the infected host [30]. Clearly, the situation in nature is more complicated than the laboratory culture; however, it is clear that mutators can and do constitute part of the natural bacterial population.

Another aspect of antibiotic resistance is what has been termed "adaptive mutation". In these cases, mutation in a subpopulation of stressed or starved cells leads to an increased rate of mutagenesis [34,35]. Recent evidence has been presented that the genome wide hypermutation may be regulated by the SOS response in response to stress [36]. Evidence has been presented for a role of the SOS induced *dinB* gene, which encodes DNA polymerase IV, an error prone polymerase, in adaptive mutation [37]. Thus the frequency of genetic variation can be increased under stress conditions such as antibiotic inhibition, nutrient starvation, or stationary phase. This in turn can lead to increased rates of resistance development as a result of the antibiotic exposure or other stresses.

It should be very clear from the preceding information that antibiotic resistance is a phenotype that can be the result of multiple and varied genetic mechanisms. Strategies for control of resistance will be far more challenging than simply instituting a more careful approach to antibiotic use. Although forced reduction in the use of a specific antibiotic can decrease the incidence of resistance, drug susceptibility may not necessarily revert to preantibiotic levels [6]. In fact, paradoxical cases of antibiotic resistance increase following reduction in utilization of a given antibiotic have been reported [6,38]. Although hundreds of antibiotics are on the market, these represent derivatives of only a small handful of structural classes. Most of these are, as mentioned previously, modified semisynthetic compounds from natural product starting points, usually derived from bacterial sources in nature. Preexisting resistance genes, presumably from the producer soil organisms, have been mobilized on promiscuous genetic elements (plasmids, phages, transposons) [39]. These widespread resistance determinants can act in some cases in concert with efflux mechanisms and mutational resistance to further increase the level of resistance. In the case of totally synthetic antimicrobials such as the fluoroquinolones and oxazolidinones, resistance has been selected that has been almost exclusively due to antibiotic target site mutations and/or efflux. It is clear that multiple gene acquisitions, mutations, and amplifications can all lead to the same phenotypic consequence, namely resistance. A comprehensive strategy to maintain control of microbial infections will require new classes of antimicrobials as well as a clearer understanding of the forces at work that promote resistance.

MICROBIAL GENOMICS AND DRUG DISCOVERY

Older antibiotic identification strategies that were successful at the dawn of the antibiotic age have not yielded additional useful classes of compounds in recent years. As a result, novel ways to identify new bacterial targets for the discovery or design of novel inhibitors have been sought. One aspect has been the continuing attempt to catalog the entire metabolic machinery of microbes to identify essential functions that might serve as starting points for the search for novel inhibitors. This program was stimulated in 1995 by the unexpectedly early arrival of the era of microbial genomics, with the delivery of the *Haemophilus influenzae* genome sequence [40]. It was widely believed up to that point that the first full sequence of a bacterial chromosome to be delivered would be that of *Escherichia coli* K-12, which was largely performed in an ordered sequence from overlapping lambda phage clones physically mapped onto the chromosome [41]. The focus on the *E. coli* genome was understandable, as a vast amount of mutant gene mapping and microbial physiology had been performed in this organism, and as a result, many genes had reliable functional annotations [42].

In contrast to the ordered approach of physical mapping and serial sequencing, the concept of "shotgun" random sequencing strategy of genomes, followed by computer assembly of the short sequences into larger "contigs" was employed with stunning success. The contigs were subsequently linked in a finishing strategy to yield a complete closed genome. The first genome completed, *H. influenzae*, was a demonstration of the power of this technique [40]. It is critical to emphasize the role of high-speed computers and sophisticated assembly software in the random shotgun sequencing approach. The overwhelming success of this approach led to its rapid adoption as the method of choice for genome sequencing. It was subsequently applied to a large number of procaryotic genomes, as well as to larger genomes [43,44]. The accumulation of genome data has occurred at a truly breathtaking rate and has been further accelerated by the introduction of a new generation of capillary electrophoresis-based DNA sequencers with increased capacity and ease of use [45]. The ability to compare, annotate, and design experiments across multiple microbial genomes via computational methods has had an influence on the very conduct of microbiological experimentation. In many cases, computer analysis of genomic data is used to generate multiple hypotheses, which subsequently are rapidly tested at the lab bench. This is a profound change from the experimental model of the recent past, in which an individual DNA sequence was obtained from a gene or region cloned as a result of a phenotype of interest.

Several pharmaceutical firms quickly understood that this sudden influx of microbial genome data could be mined for sets of novel targets for both antibiotic and vaccine development [45]. In addition to identifying a significant number of novel targets, the microbial genome efforts serve as a test bed for development of advanced computational tools and pioneering genomic strategies such as DNA microarrays for expression analysis [46,47,48,49] and proteomic analysis of cells [50,51]. Numerous innovative, but in many cases unproven, drug discovery tools have been generated mainly by small biotechnology firms, and antibiotic discovery has been an early focus of several of these efforts. The microbial genomics experience, dealing with smaller genomes and organisms that have readily accessible genetic systems, are in preparation for the larger assault on identifying targets from the human genome and developing drugs for these targets. In the case of microbes, the drugs sought are selective toxins; in the mammalian case, agonists and antagonists of specific cell targets are sought to modify metabolism.

NOVEL TARGET IDENTIFICATION

The underlying assumption of employing this technology is that microbial genomics will provide a number of essential, validated targets as possible candidates for the discovery of new antimicrobials. The ideal new antibacterial target should include the following properties: (1) It should be broad spectrum in that it should be present in multiple pathogenic bacteria. Proteins of high sequence similarity that are distributed among microbes, termed "orthologs", are considered high-value targets. (2) The gene would encode an essential function whose inhibition would result in a bacteriostatic or bactericidal effect on the pathogen. (3) An inhibitor of the target would be a novel chemotype lacking cross-resistance with currently used clinical agents. (4) Selectivity for bacterial cells over eukaryotic cells in order to reduce the likelihood of adverse effects that would be associated with a novel compound in the clinic. To evaluate targets for such criteria, a combination of bioinformatics and experimental biology is required.

There are two fundamentally different philosophies that may be taken to the identification of novel targets in bacteria. In the first approach, genes that are essential for in vitro growth (i.e., growth on artificial bacteriological media) are identified by several methods, some of which are discussed in detail below. There is strong precedence for inhibition of in vitro essential targets as excellent antimicrobial targets. Virtually all the antibiotics presently in use in the clinic inhibit bacterial growth on or in artificial media, and this in fact forms the basis for current clinical antimicrobial susceptibility testing [52]. A different, novel strategy is to identify targets that are only expressed or only essential in the context of the infected host. Inhibitors of these processes would presumably interfere with the infection process and/or continued survival of the microorganism within infected tissues and would constitute a previously undiscovered class of antimicrobial inhibitors [53]. These targets would constitute an "in vivo" essential target class.

Essential processes that are potential antimicrobial targets can be identified by two broad strategies. In the first case, essential genes may be identified in the laboratory by what are nontargeted or essentially random methods of gene disruption. In these cases, large populations of organisms are mutagenized by one of several methods, and the resulting subset of cells is analyzed for mutants that would define specific targets. In contrast, a second strategy consists of specific gene targets that are preselected on some basis and disrupted for essentiality testing in a target-specific fashion.

ESSENTIAL GENE IDENTIFICATION BY RANDOM MUTAGENESIS

One example of the first process of random target generation was the identification of a set of conditional mutants which all possessed a phenotype that is nonpermissive for growth at extreme temperature, termed ts (temperature sensitive) mutants [54]. The temperature-sensitive genes presumably produce thermolabile proteins that, because of the nonpermissive phenotype, are generally believed to be essential to survival. In this case, a culture of bacterial cells is usually mutagenized and screened for mutant progeny with the temperature-sensitive characteristic. Specific gene mutations are subsequently identified, usually by screening a plasmid library constructed from the parental wild-type genome for the restoration of growth at nonpermissive temperatures. The complementing gene carried by a plasmid is then sequenced to identify the probable gene identity, and confirmation by a number of standard genetic techniques of the temperature-sensitive essential gene is performed [55]. An advantage to this approach is that the temperature sensitive strains can be readily employed in high-throughput screens to search for inhibitors. The supposition is that even at the growth permissive temperature, the thermolabile essential protein is still suboptimal, and the cells would therefore be more sensitive to an inhibitor of that target protein. In this case, as in the case of the transposon mutagenesis, described below, the bioinformatics analysis is carried out after the identification of the gene responsible for the conditional phenotype. Obviously, it is a great advantage if the organism selected for the process is one whose genome has already been sequenced.

Another random methodology to identify potential essential genes that could constitute targets is the use of transposon mutagenesis [56]. After introduction into a culture of bacterial cells, the transposons will insert into multiple genomic locations in the different cells of the population. In this case, the transposons can act as gene knockout systems by insertion into the genome and disrupting control or coding sequences. The transposons usually bear some form of selective marker, such as antibiotic resistance, to readily identify the transposon-bearing cells. Since transposons disrupt genes into which they insert, they can be used to identify essential genes by a process of elimination. Failure to repeatedly observe a transposon insert into a region of the genome would be indicative of an essential function in that region [57,58]. Thus, the transposon strategy to detect essential genes is in essence a negative one, in which the lack of an insertion event is sought. A drawback with all transposon-based gene essentiality tests is the sheer numbers of insertions necessary to saturate the chromosome with statistically significant numbers of transposon inserts. The transposon insertion sites may be localized in the genome by DNA sequencing using a known transposon primer sequence extended into the adjacent genes. Alternatively, a footprinting procedure may be used to locate transposon insertions near and within a gene of interest. Insertion of a transposon into an essential gene will lead to the loss of that cell and its progeny from the pool. To screen for this loss, a PCR primer within the transposon is used along with an opposing primer near the region of interest. The resulting PCR products can be sized on a gel, and failure to recover viable transposon insertions in a gene would result in a "gap" in the PCR band sizes corresponding to the gene. The gene would therefore be presumed to be essential. Another method, utilized in yeast, follows the loss of strains with inserts into essential genes over time from a population mutagenized with transposons [59].

The transposon disruption process can be confounded, however, due to the preference for certain insertion site sequences ("hot spots") exhibited by most transposons [60]. This problem of nonrandom inserts has been largely alleviated by the adaptation of an in vitro tranposon system based on the *mariner* transposon of the horn fly [61]. This system uses a modified *mariner* element (antibiotic resistance gene flanked by the *mariner* inverted repeats), the transposase, and purified bacterial chromosomal DNA as a target [62]. The *mariner* system exhibits little site preference, needing only the dinucleotide TA in the target sequence. The strategy, termed Genome Analysis and Mapping By *In vitro* Transposition (GAMBIT) reintroduces the pools of in vitro transposons into a transformable bacterial species (e.g., *H. influenzae* or *S. pneumoniae*). Genetic footprinting, as described above, is used to determine regions in which the target genes are designated essential by virtue of the lack of recoverable inserts. This system has been further developed into a transposon termed TnAraOut, which uses the *mariner*-based transposition and an outward-facing arabinose-inducible promoter P_{BAD} [63]. In this case, transposition into a promoter region of an essential gene results in an arabinose-dependent cell growth phenotype, dependent on the inserted arabinose promoter expression. This allows the essential genes so identified to be regulated by the arabinose concentration in the media. A number of known essential as well as previously undiscovered essential genes were uncovered with this system.

Transposon systems can also be readily adapted to search for genes that may only be expressed in the infected host and be important to the development or progression of the infection process. This could lead to the identification of inhibitors that work in a different manner than traditional antibiotics. A specific example of this is the Tn*phoA* system and its use to identify genes that encode virulence related surface proteins for *Salmonella typhimurium* and *Vibrio cholerae* [64]. It is conceivable that a small molecule that interfered with the function of this surface protein would attenuate the intestinal infection process of these two pathogens.

Finally, it should be noted that another problem, unique to procaryotes, also complicates transposon-based essential gene hunting. This is the fact that many bacterial genes exist in operons, which are transcriptionally linked genes, often comprising a metabolic pathway [65]. Disruption of an upstream gene in this cluster by a transposon or other insert can profoundly affect downstream gene expression; a situation termed polarity [56,66]. Computational and experimental analysis of the genome in the region of a putative essential gene is necessary to either exclude polarity or to design further experiments to address the individual genes in an operon structure.

There are several additional methods that rely on nontargeted, random genetic manipulations to discover essential processes. One such technique, cassette mutagenesis, uses restriction enzymes to digest chromosomal DNA in vitro; circularize the DNA; cut with a second enzyme and insert an antibiotic resistance cartridge, thus recircularizing; and finally digest with the original restriction endonuclease to linearize the DNA for introduction via transformation [67,68]. Selection for the antibiotic resistance identifies clones that have received cassette inserts and still retain viability. By a process of insert saturation of the genome and the failure to detect inserts in certain genes, essentiality may be inferred. This technique has been used to investigate genes essential to genetic transformation in *H. influenzae* [68].

Another strategy that relies on a random process and subsequent phenotype screening is the use of randomly generated antisense RNA fragments [69]. In this approach, sheared chromosomal DNA is inserted into a tetracycline-regulated expression vector and introduced into *S. aureus* and expression induced by the weak antibiotic anhydrotetracycline. Inserts were identified that led to reduction or halting of growth. Approximately one-third of the growth inhibitory DNA fragments were found to be antisense-oriented gene fragments. The exercise revealed more than 150 critical staphylococcal genes where antisense expression led to growth inhibition.

In the above situations, where random gene disruptions are created, the bioinformatic analyses of the data is performed after the initial "wet" biology. Locating the position of a temperature-sensitive mutant or transposon in a sequenced genome is followed by an informatic analysis of the altered gene and its surroundings. In the case of temperature-sensitive mutants, comparison of the Ts form of the gene with the wild-type sequence will disclose valuable information on the location of the mutation within the gene and possibly the function of the gene. If the gene is in a partially characterized operon, further functional information may be gleaned. In the case of tranposon insert libraries, informatic analysis will reveal which genes have multiple transposon inserts and which genes do not harbor inserts in situations of saturation mutagenesis of a genome. The noninsert "holes" would signal the presence of potential essential genes. The computational analysis of the surrounding region would indicate the likelihood of operon structure and the need for further biological experimentation. Computational comparisons will also reveal the distribution of orthologs of potential essential proteins among pathogenic species of interest.

ESSENTIAL GENE IDENTIFICATION BY TARGETED GENE KNOCKOUTS

A second strategy to identify essential genes uses genomic information to specifically select genes for subsequent targeted disruption in the chromosome. In this case, the initial informatics analysis is done up front to determine which genes should be targeted based on specific criteria. For example, genes can be grouped based on their distribution among a variety of gram-negative and gram-positive bacteria at a given level of amino acid similarity of their gene products. The assumption is that orthologous proteins with highly conserved amino acid sequences will have similar three-dimensional structures (and, by extension, similar function) than proteins that do not share similarity. Targets with the highest similarity among several pathogens can be assumed to offer the best hope for broad specificity. Alternatively, genes specific to either gram-negative or gram-positive bacteria can be identified. There are several successful currently marketed antibacterial agents with such specificities, albeit not based on genomic differences [70,71]. This approach can also be used to find organism-specific targets that can lead to the discovery of "niche" drugs. For example, pathogens such as the mycobacteria are rich in potential targets not present in other bacteria. These niche antimicrobials would be of interest for organisms such as *Helicobacter pylori* or *Mycobacterium tuberculosis* [72]. Using a similar reasoning, comparison of the targets selected with human genomic and other eucaryotic databases can provide an indication of possible selectivity of inhibitors, thus reducing the progression of mechanism-based toxic compounds.

COMPUTATIONAL-BASED TARGET SELECTION

An example of one such bioinformatic application for target selection was described in a Concordance analysis of microbial genomes [73]. This system performs a FASTA comparison of multiple genomes at the amino acid level and builds tables for subsequent data access by a web-based interface. The retrieval of sequences can be made based on end-user-specified similarities and organism selections, which distinguishes it from other comparison tools such as COG, HOBACGEN, and others which use fixed default constraints [74,75]. The Concordance published compared the *E. coli* genome against *B. subtilis*, *H. influenzae*, *H. pylori*, and *M. tuberculosis* using a BLOSUM62 matrix and subtracted out eukaryotic sequences with similarity greater than a selected exclusion criterion using the yeast *Saccharomyces cerevisiae*. Almost 2000 sequences were found with a match to at least one of the five species under the selected criterion; however, only 265 matched all species. The final result was the selection of 89 sequences in common with all five bacterial species with an additional 176 sequences eliminated because of similarity to yeast sequences. The Concordance will display CLUSTALW multiple alignments of the protein similarities among organisms upon user request. The utility of such an approach was demonstrated by examination of the sequences ultimately selected. For example, the *gyrA* gene encoding DNA gyrase, the target of the quinolone class of antibiotics was identified as well as *murA*, the target for fosfomycin. Several additional previously reported essential genes were found, including *dnaA*, *ftsZ*, and *mraY*. Different organisms and user-specified criteria can be used for such analyses depending on the desired endpoint. Also, organism-specific sequences can be discovered with this approach.

Another program that accomplishes a similar subtractive analysis is Find Target [76]. In this Unix-based system, a BLASTP comparison of proteomes is accomplished based on user specified parameters, although the program can also use FASTA or PSI-BLAST for similarity. The program has several utilities for the researcher to examine the details of each match, such as global alignments and phylogenetic trees.

Use of the above programs, as well as others [77–79], can define sets of microbial genes that can be tested for fitness as novel antimicrobial targets. As discussed previously, criteria might include broad distribution of the putative targets among pathogenic bacteria and minimal similarity to available eucaryotic genomes, most notably the human genome. At this point, the computational programs used to sort proteins as to their similarity on genomewide scales rely on primary amino acid relationships. These programs are used to generate lists of genes for essentiality testing. An important point for consideration is that limited sequence similarity should not be used alone to evaluate targets. There are several examples available to illustrate the risk of using limited sequence similarities to identify potential targets. Examples include human and *E. coli* dihydrofolate reductase, the antibacterial target for trimethoprim (28% identical at the amino acid level), and human and *E. coli* topoisomerase II, the target for quinolones (20% identical). From a different perspective, subfamilies of the penicillin binding proteins (PBPs) show low overall identity at the amino acid level to each other, but inhibitors often show broad specificity with strong affinities for PBPs across numerous bacterial species [80]. This is due to tertiary structural similarities and small conserved regions that make up both the active site and the site of inhibition by β-lactam antibacterial compounds. Therefore, it is possible to have selectivity despite similarities to human orthologs or broad-spectrum targets even though the level of overall identity of proteins among various bacterial species appears low.

Another consideration in attempting to identify essential genes by targeted gene disruption is that although one is preselecting targets with broad bacterial distributions, it is both technically and pragmatically impossible to test these targets in multiple bacterial species. Many pathogens do not readily lend themselves to efficient gene disruption, although, as will become clear in the following section, efforts to develop and broaden genetic tools are progressing. The manpower costs of performing a large-scale gene knockout campaign also severely limit the number of individual pathogens that can be interrogated with regard to gene essentiality. In the following section, genetic systems that are adaptable to the task of high-throughput examination of gene essentiality in several bacterial systems is examined.

TARGETED GENE KNOCKOUT SYSTEMS

As previously discussed, *E. coli* represents the best characterized organism in terms of genes and metabolism. This is due largely to the vast number of mutants that have been isolated and defined over 50 years of intense study. The analysis was aided by genetic manipulations (conjugation, phage transduction, transposition) that permitted the construction of a detailed physical map [81] and a significant proportion of individually sequenced genes. The publication of the complete genome sequence marked the complete delineation of the genes of *E. coli* K-12 [41]. It is sobering, however, that despite the intense genetic and physiological work performed over a half-century, almost 40% of the open reading frames detected were initially annotated as having unknown function. Nonetheless, much of the annotation of genome sequence information of other microbes is based on comparison to *E. coli* orthologs. It was only recently, however, that *E. coli* genetic systems were devised to perform gene knockouts in a relatively straightforward fashion.

One knockout system, pKO3, was based on a derivative of an earlier plasmid, pMAK 700 [82], that was temperature sensitive for replication. The pKO3 plasmid has both positive (chloramphenicol resistance) and negative (sucrose sensitivity) selection and can deliver PCR constructs into the chromosome via homologous recombination [83]. Subsequent resolution of the integrated construct is performed at 30°C, and selection for plasmid loss is in high sucrose, with loss leading to restoration of chloramphenicol sensitivity. By careful construction of PCR products, the system can be used to create precise deletions or insertions into the *E. coli* chromosome. A demonstration of the utility of the pKO3 plasmid was the identification of six *E. coli* unknown function essential genes identified among 26 genes that were conserved in *Mycoplasma genitalium* [84]. This system was also recently employed to identify six novel, essential genes in *E. coli* among a set of 27 genes conserved in *H. influenzae*, *S. aureus*, *S. pneumoniae*, and *Enterococcus faecalis* [85]. The drawback of all the plasmid-based systems is that multiple steps are involved in integration and resolution of the plasmids. One strength is that by careful construction, precise gene deletion is possible, thereby minimizing potential polar effects in operons.

Methods for introduction of linear DNA segments into the *E. coli* chromosome have also been perfected in recent years. It has long been recognized that *E. coli* contains exonucleases that degrade linear DNA that is exogenously introduced [86]. Early attempts to circumvent this problem include the use of *recBC*, *sbcB* mutants, which are inactivated in exonuclease V or *recD* mutants, which are similarly defective in exonuclease V [87,88]. These methods yielded low numbers of recombinants and required extensive homologies between the introduced DNA and the target chromosomal location. More recent developments have employed the λ phage Red genes (γ, β, and *exo*) expressed from either a chromosomal, low copy plasmid or defective λ prophage construct [89–91]. The λ Red system is part of the phage recombination system, and upon expression, promotes a greatly enhanced rate of recombination. In addition, the *gam* (γ) gene inhibits exonuclease V. The overall result is *E. coli* cells that can take up and integrate linear DNA after electroporation. These systems have been demonstrated to work with short regions of target region homology (ca. 50 bp) flanking a heterologous DNA insert (e.g., antibiotic resistance gene) [91]. The short regions of homology may be incorporated synthetically into the PCR primers used to amplify the heterologous insert (see Figure 1). Systems such as these offer the real possibility of a rapid, systematic approach to gene replacement and deletion in *E. coli.* For example, one of the reports on the Red system made 40 different

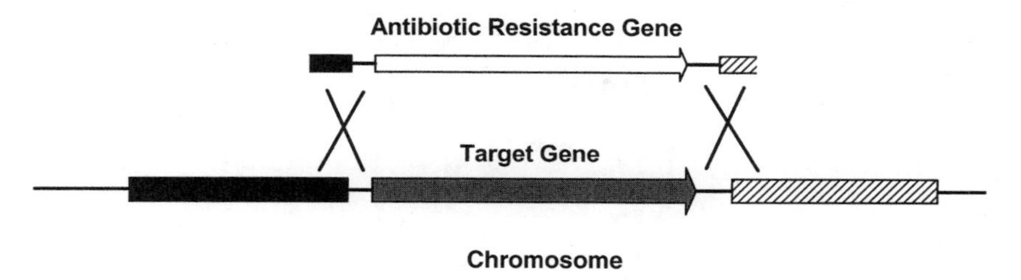

Figure 1 Replacement of a target gene by a linear product with flanking sequence homology. This generalized diagram illustrates the principle of heterologous gene replacement via a double crossover promoted by flanking homology within a linear DNA product. In the case of *E. coli* λ *red*, *gam* system described in the text, the flanking regions can be less than 50 basepairs on either side of the heterologous gene. These flanking regions of homology may be incorporated into the PCR primers used to amplify the antibiotic resistance gene. Recovery of an antibiotic resistant strain via integration of the resistance gene would indicate that the target gene was not essential for cell survival.

disruptions in the *E. coli* chromosome, using 36-bp flanking homologies to the targets on either side of a resistance gene [90].

With regard to gram-positive microorganisms, genomic sequencing of several key organisms has stimulated the development of genetic tools to probe gene essentiality, function, and structure. The gram-positive counterpart of *E. coli* with regard to the most information developed in genetics and physiology is undoubtedly *Bacillus subtilis*. Plasmid vectors for *B. subtilis* that are specifically designed to address multiple questions are the pMUTIN series [92]. The pMUTIN plasmids, with an *E. coli*-based Co1E1 replicon, cannot replicate in *B. subtilis*. The plasmids contain a β-lactamase gene for plasmid manipulations in *E. coli*, an erythromycin resistance expressed in gram-positives, and a multiple cloning site to insert *B. subtilis* DNA that, by homologous recombination, will promote single crossover integration of the plasmid via a Campbell-like mechanism (see Figure 2A) [93]. By inserting a central DNA fragment (lacking substantial parts of the N- and C-terminal coding regions) of the gene targeted for disruption, the plasmid recombines into the chromosome, yielding a partial insertion duplication of the target gene flanking the integrated plasmid. If the target gene is nonessential to in vitro survival, erythromycin-resistant cells will survive and colonies will be recovered. If a gene is essential, no erythromycin-resistant progeny will be observed. The pMUTIN-4 plasmid also contains the pSpac promoter, with an altered operator and strong upstream transcriptional terminators to permit tight regulation of a gene from the integrated pSpac. By inserting the 5′ end of the

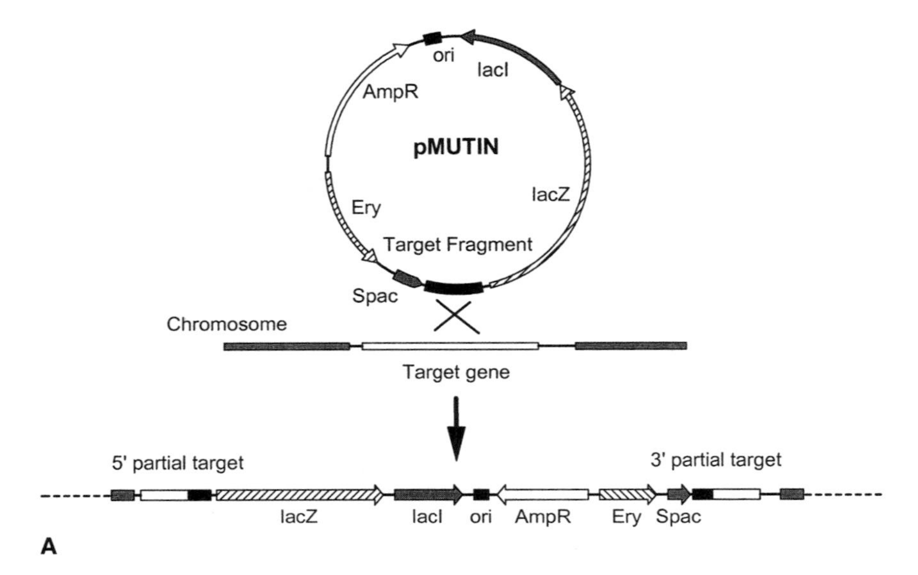

FIGURE 2A Integration of pMUTIN into the Bacillus subtilis chromosome. This plasmid, which cannot replicate in *B. subtilis*, was specifically designed for high-throughput knockout and gene analysis in *B. subtilis*. In the protocol outlined above, the plasmid is being used for a targeted gene knockout. The central region of the target gene, lacking significant sequence from both the 5′ and 3′ regions of the gene, is amplified via appropriate PCR primers. This fragment is inserted into pMUTIN and the resulting plasmid introduced into *B. subtilis*. The fragment acts as a region of homology, facilitating an insertion of the plasmid via a Campbell-like mechanism, flanked by partial, nonfunctional target gene duplications. Recovery of transformed colonies that are erythromycin resistant indicates that the disruption of the target gene was not lethal. The promoterless *lacZ* gene can be used to monitor native promoter strength under different culture conditions.

target gene behind the Spac promoter, the resulting integrant will generate a complete copy of the target gene under Spac control (see Figure 2B). This feature also allows an analysis of potential polarity effects caused by the insertion of the pMUTIN plasmid into an operon. The pSpac promoter makes it possible to express genes "downstream" of the plasmid insert in a potential operon.

Community acquired respiratory disease is a major medical problem, and *Streptococcus pneumoniae* is a major gram-positive pathogen responsible for the majority of community-acquired pneumonia. Pneumococci are readily transformable by exogenous DNA when made competent, a previously somewhat difficult process recently rendered straightforward by the discovery of a competence-

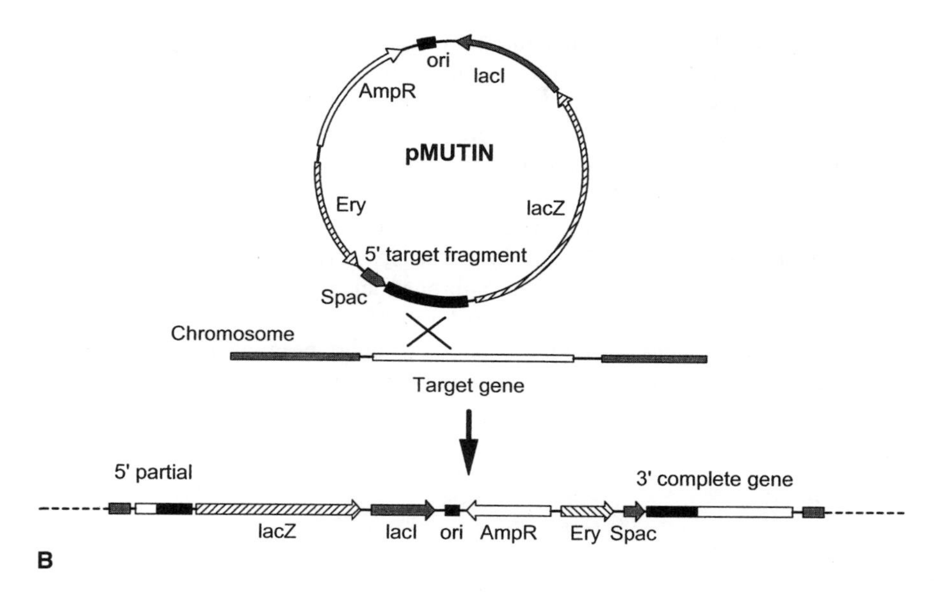

FIGURE 2B pMUTIN integration and transcriptional control by the Spac promoter. By using a target fragment that contains only the 5′ end of the gene and is oriented such that expression can be driven by the Spac promoter, an insertion can be generated that yields a partial gene duplication, the plasmid insertion, and a complete gene behind the Spac promoter. Since Spac can be regulated by exogeneous IPTG, the target gene becomes dependent on the level of IPTG for expression. In this manner, essential genes can be recovered with colony formation conditional upon IPTG presence.

inducing peptide [94]. In addition, the pneumococcus contains active recombination pathways, making these organisms ideal candidates for gene knockout experiments. A plasmid to create gene disruptions by homologous recombination leading to insertion duplication in pneumococci was designated pEVP3 [95]. Similar to the later pMUTIN system, pEVP3 replicates only in a gram-negative (*E. coli*) background. The plasmid carries chloramphenicol resistance, which is expressed in both gram-positive and gram-negative backgrounds. Insertion of a central fragment of the target gene for knockout generates a plasmid capable of insertion–duplication via recombination, although due to the mode of pneumococcal DNA uptake, the process is not a straightforward Campbell integration. The result, however, is that the target gene is disrupted. If the target is nonessential, chloramphenicol-resistant colonies will be recovered; if essential, no colonies will result on chloramphenicol-containing agar plates. Another plasmid for use in the pneumococci, pRKO2, has a similar mechanism of gene disruption by insertion of the plasmid promoted by homologous recombination of a gene fragment [96]. This plasmid also contains an optimized tetracycline promoter, which permits controlled expression of the gene where the insert has occurred. Demonstration of the utility was made by integrating the plasmid into the known essential *gyrA* gene, placing the gene under the control of the tetracycline promoter. The resulting pneumococcal strain was found to be dependent on tetracycline for growth. Development of pneumococcal plasmid knockout systems has resulted in the ability to perform gene knockouts in a high-throughput manner, identifying multiple new targets for potential antibiotic development [97].

Systems for gene disruption in another important gram-positive pathogen, *Staphylococcus auneus*, have also been developed (see Figure 3). One disruption

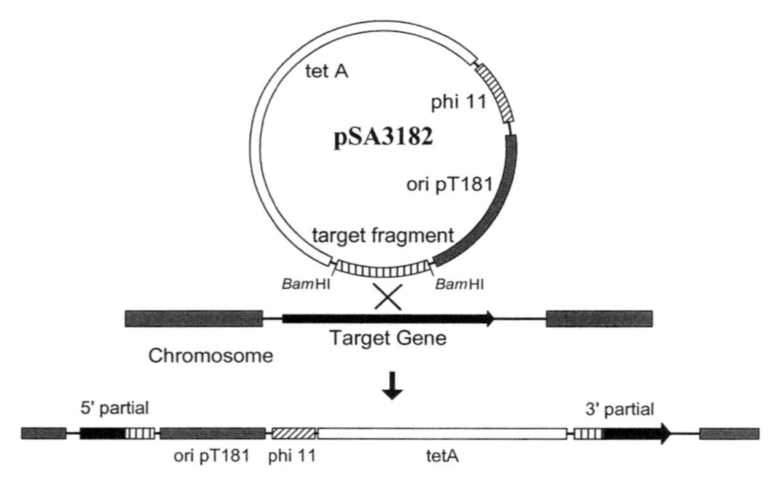

FIGURE 3 Insertion of plasmid pSA3182 into the Staphylococcus aureus chromosome via a Campbell-like insertion duplication. A fragment of the target gene lacking the 5′ and 3′ coding regions is inserted into pSA3182 at a unique *Bam*HI site. Upon introduction into an *S. aureus* strain lacking RepC, the plasmid cannot replicate and recombines into the chromosome at the target gene site, leading to insertion duplication of partial gene fragments at either end of the inserted plasmid. Neither copy of the gene that flanks the inserted plasmid is full length, and, therefore, both are nonfunctional. If the gene is nonessential, a tetracycline-resistant cell, due to the integrated *tetA* gene, will result. The plasmid is initially replicated in a strain of *S. aureus* RN4220 that has multiple copies of the *repC* gene on transposon inserts. The presence of the ø11 gene permits packaging by the phage and transduction of the plasmid into the target *S. aureus* strain at very high frequences (>5%).

vector, pSA3182, carries tetracycline resistance and is dependent on the presence of the *repC* gene for autonomous replication [98]. Strains of *S. aureus* that contain RepC will permit maintenance of the plasmid, whereas strains lacking the RepC protein will retain tetracycline resistance only if the plasmid is integrated into the chromosome. As in the above examples, this integration is dependent on the homologous recombination of the plasmid by single crossover into the chromosome, promoted by the target gene DNA fragment. Failure to observe tetracycline-resistant recombinants indicates that the disrupted gene that was targeted was essential. The plasmid also contains a segment of the ϕ11 phage, which permits phage packaging and transduction of the plasmid among staphylococcal strains. It is possible, by PCR, to construct a gene with a central deletion in it. Introduction of this construct into the plasmid will result in the insertion duplication of the deleted gene upon plasmid integration. Subsequent introduction of a replicative plasmid that overproduces RepC leads to loss of the integrated pSA3182. The excision can resolve such that the gene deletion remains in the chromosome, yielding an unmarked deletion. By careful construction, the deletion can be polarity neutral, permitting expression of downstream genes.

Integrating plasmid constructs with promoters that can be exogenously regulated have also been developed for *S. aureus*. One derivative, pFF81, carries the Spac promoter from *Bacillus* phage to drive transcription of the target gene after plasmid integration into the chromosome [99]. A similar construct that also uses the Spac promoter was used to demonstrate the essentiality of the *murE* gene in *S. aureus* [100]. In this case, plasmid replication is temperature sensitive, permitting selection of antibiotic-resistant integrants into the chromosome at high temperatures. A two-step procedure that places any *S. aureus* gene of interest behind a controllable promoter has also been developed [101]. In this procedure, the gene under investigation is initially placed in a construct behind a controllable promoter such as Spac or Xy1/tet, along with a selectable antibiotic marker. This construct is integrated via a site specific into an ectopic site in a nonessential gene in the *S. aureus* chromosome. The native gene is then disrupted, and the gene under promoter control is regulated via exogenous inducers. This permits titration of the expression levels, resulting, in the case of essential genes, in inducer dependent growth.

Gene disruption via flanking homology and recombination in various bacterial systems has been facilitated by the use of PCR-based techniques such as splice overlap extension (SOE) [102,103]. This *in vitro* construction method, outlined in Figure 4, permits the precise fusion of two or more genes through the use of carefully designed PCR primers. It can also be employed to place regulated promoters in front of a gene, introduce precise point mutations, generate precise gene deletions by fusing genes on either side of a targeted gene, and many other manipulations. The ability to reintroduce such engineered fragments back into a transformable bacterial species (either natural or artificial transforma-

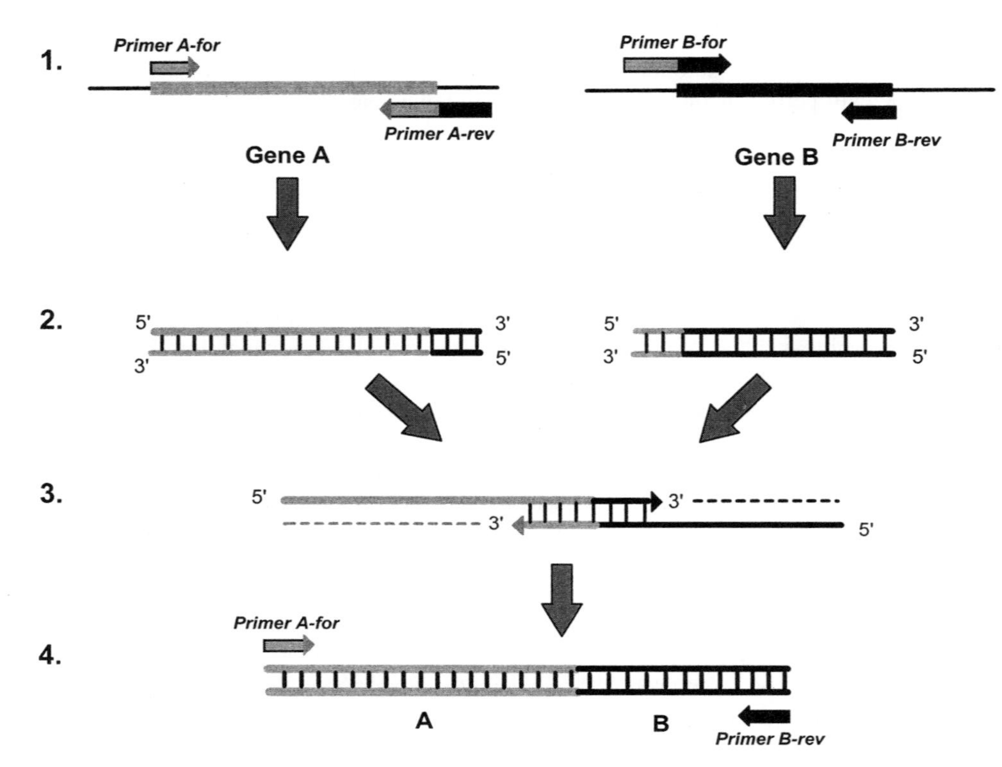

FIGURE 4 The SOE PCR technique. This in vitro method permits the construction, via design of PCR primers, of a number of different recombinant molecules. In this example, two genes are precisely spliced together in a four-step operation. In step 1, the two genes are amplified separately via PCR, using primer pairs 5′ A-forward and 3′ A-reverse and then 5′ B-forward and 3′ B-reverse. The 3′ A-reverse reverse primer has a region that is identical to the sequence of the 5′ end of gene B, and the 5′ B-forward has a region identical in sequence to the 3′ end of gene A (that is, the two "joining" primers are reverse complements of each other). In step 2, the two genes are amplified separately via multiple rounds of PCR. In step 3, the products of step 2 are mixed and annealed. The complementary strands that have free 3′ ends will be extended via rounds of PCR in this step. Finally, in step 4, the original gene A 5-forward primer and the original gene B 3′-reverse primer are used to amplify the new, longer spliced product. An analogous reaction can be used to place a controllable promoter upstream of a gene, substituting the promoter for gene A in the reaction.

tion), with subsequent integration via homologous recombination, is a powerful analytical tool in understanding gene function and regulation.

BEYOND TARGET IDENTIFICATION—GENES TO SCREENS

All of the above methods are examples of the tools that can be employed on a large scale to probe microbial genomes for genes of interest as antimicrobial targets. The identification of the genes that represent potential novel targets is a fundamental early step along the path to novel compounds. It should be appreciated that this is the first of many subsequent steps to progress a candidate gene to the point where the encoded protein can serve as a substrate for, e.g., a high-throughput screen, or for structural studies. In the case of essential genes that have significant similarities and identities to genes of known function, it is relatively straightforward to test if the proteins encoded have similar biochemical activity. In the case of conserved unknown function genes, several additional technologies (many examples of which are covered in chapters of this book) may be employed to discern possible function and/or screen for inhibitors. Potential roles of unknown function genes may be explored experimentally by DNA microarrays and proteomics to discern common regulatory patterns and physiological responses [104–107].

Bioinformatics also has a role to play at this stage, as programs that compare genes and annotate by marshaling multiple lines of evidence and programs that reconstruct metabolism can both bring value in delineating potential roles for unknown function genes [108,109]. Additional informatics analyses may include searches for sequence patterns, for example, the Pfam database tools and HMMER, which are based on Hidden Markov Models [110]. A recent comprehensive discussion of bioinformatics tools covers the above topics and additional genome analysis programs [111]. Many of these tools are based on different models and computational approaches, and judicious use can pay dividends in suggesting functional possibilities for unknown genes.

Despite the identification of a large number of essential genes conserved among a broad array of pathogens using genomics approaches, a key question is the overall "quality" of the targets. One dimension of a target's quality is undoubtedly how rapidly resistance will arise. In this regard, antimicrobials that target multiple, physiologically related targets within a bacterial cell might be less susceptible to rapid high-level resistance development. Several examples exist of current antibacterial drugs that target multiple, functionally related targets within bacteria. The targets of the β-lactam antibiotics are multiple penicillin binding proteins, which participate in peptidoglycan synthesis and other essential cellular processes [112]. Likewise the quinolone antibiotics have dual targets, two enzymes (gyrase and topoisomerase IV) that maintain the proper supercoiling and decatenation of the bacterial chromosome [113]. Mutational resistance to these targets tends to be incremental stepwise changes. In contrast, rifampin, which targets a single protein, RNA polymerase, is noted for the relatively rapid development of single step, high-level resistance [114,115]. If the presence of multiple targets within a cell is deemed important to reduce resistance development, it may be necessary to investigate functionally related proteins (paralogs) within species as potential sources of multiple targets. Other target-quality considerations are less well understood and formulated. To identify inhibitors, it will be necessary to either randomly screen the targets against libraries of small molecules or obtain detailed three-dimensional information on their active sites for attempts at de novo inhibitor design. In either case, it may lead to a target inhibitor that has little activity in whole cells. Unfortunately, the chemical features that permit penetration of inhibitor molecules into bacteria are critical but poorly understood properties [116]. The problem extends further as our abilities to craft an inhibitor molecule that can both penetrate to a target and fit into a critical binding pocket, while simultaneously possessing the necessary human pharmacological properties to be a drug, are currently fundamentally lacking. While the proximate application of genomics has been successful in microbial target identification, there remain nonetheless significant obstacles to translation of this information into therapeutic agents. Additional work in microbial membrane permeation properties and further refining the pharmacological parameters that make a molecule "druglike" are necessary to eventual success.

EPILOGUE

It is certain that with the discovery, development and clinical use of novel antibacterial classes, there will also come new mechanisms of resistance. To limit such development, it would be tempting to reserve novel compounds for use solely in cases of severe infection from multiresistant organisms. However, this argument has two flaws: First, existing drug resistance is not a static situation; survey data indicate that it continues to increase overall in both the hospital and community settings [117]. It is currently true that many infections eventually respond to some form of treatment. However, it can be anticipated that there will be continued erosion in treatment options due to resistance. This in turn will force the increased use of novel agents as they become the only option for multiresistant organisms. Second, to insure the continued development of novel agents, it is necessary to create incentives for industry to develop them. Arbitrary and limited use and reimbursement policies will further discourage investment in antimicrobial development. Overall, it is more a choice of using the right antibacterial rather than an "old" or "new" one in a given clinical situation. Clearly, some global consensus that takes into account both the medical and economic consequences of the development and use of novel agents should be debated, defined, and enunciated.

Success at formulating novel antimicrobials will give us a "second chance" at effective infection control. It will be important to use this resource wisely.

REFERENCES

1. J Lederberg. *Science 288*:287–293, 2000.
2. L Silver, K Bostian. *Eur. J. Clin. Microbiol. Infect. Dis. 9*:455–461, 1990.
3. LL Cavalli, GA Maccaro. *Nature 166*:991–992, 1950.
4. J Davies. *Science 264*:375–382, 1994.
5. MA Espinal, A Laszlo, L Simonsen, F Boulahbal, SJ Kim, A Reniero, S Hoffner, HL Rieder, N Binkin, C Dye, R Williams, MC Raviglione. *N. Engl. J. Med. 344(17)*: 1294–1303, 2001.
6. S Monroe, R Polk. *Curr. Opin. Microbiol. 3*:496–501, 2000.
7. IM Gould. *J. Antimicrob. Chemother. 43*:459–465, 1999.
8. HC Neu. *Science 257*:1064–1073, 1992.
9. SB Levy. *N. Engl. J. Med. 338*:1376–1378, 1998.
10. DT Moir, KJ Shaw, RS Hare, GF Vovis. *Antimicrob. Agents Chemother. 43*: 439–446, 1999.
11. C Walsh. *Nature 406*:775–781, 2000.
12. DC Hooper. *Drug Resist. Updates 2*:38–55, 1999.
13. K Bush. *Clin Infect. Dis. 32*:1085–1089, 2001.
14. DJ Diekema, RN Jones. *Drugs 59*:7–16, 2000.
15. IJ Mare. *Nature 220*:1046–1047, 1968.
16. R Gomez-Lus. *Int. Microbiol. 1*:279–284, 1998.
17. MCJ Maiden. *Clin. Infect. Dis. 27*:S12–S20, 1998.
18. HFJ Chambers. *Infect. Dis. 179*:S353–S359, 1999.
19. A Phillippon, R Labia, GA Jacoby. *Antimicrob Agents Chemother. 33*:1131–1136, 1989.
20. S Handwerger, MJ Pucci, KJ Volk, J Liu, MS Lee. *J. Bacteriol. 174*:5982–5984, 1992.
21. C Walsh, SL Fisher, IS Park, M Prahalad, Z Wu. *Chem. Biol. 3*:21–28, 1996.
22. MI Borges-Walmsley, AR Walmsley. *Trends Microbiol. 9*:71–79, 2001.
23. N Bilgin, F Claesens, H Pahverk, M Ehrenberg. *J. Mol. Biol. 224*:1011–1027. 1992.
24. RA Fisher. *The Genetical Theory of Natural Selection*. New York: Oxford University Press, 1930.
25. J Björkman, DI Andersson. *Drug Resist. Update 3*:237–245, 2000.
26. J Björkman, I Nagaev, OG Berg, D Hughes, DI Andersson. *Science 287*:1479–1482, 2000.
27. B Levin, V Perrot, N Walker. *Genetics 154*:985–997, 2000.
28. F Taddei, M Radman, J Maynard-Smith, B Toupance, PH Gouyon, B Godelle. *Nature 387*:700–702, 1997.
29. F Baquero, M-C Negri, MI Morosini, J Basquez. *Clin. Infect. Dis. 27(suppl. 2)*: S5–S11, 1998.
30. JE LeClerc, B Li, WL Payne, TA Cebula. *Science 274*:1208–1211, 2000.
31. JL Martinez, F Baquero. *Antimicrob. Agents Chemother. 44*:1771–1777, 2000.

32. P Funchain, A Yeung, JL Stewart, R Lin, MM Slupska, JH Miller. *Genetics 154*: 959–970, 2000.
33. A Oliver, R Canton, P Campo, F Baquero, J Blazquez. *Science 288*:1251–1254, 2000.
34. PL Foster. *Ann. Rev. Genet. 33*:57–88, 1999.
35. SM Rosenberg. *Curr. Opin. Genet. Dev. 7*:829–834, 1997.
36. GJ McKenzie, RS Harris, PL Lee, SM Rosenberg. *Proc. Natl. Acad. Sci. USA 97*: 6646–6651, 2000.
37. GJ McKenzie, PL Lee, M-J Lombardo, PJ Hastings, SM Rosenberg. *Mol. Cell 7*: 571–579, 2001.
38. Division of STD, Dept. of Health and Human Services. *Sexually Transmitted Disease Surveillance 1998 Suppl. Gonococcal Isolate Surveillance Project*. Atlanta: CDC, 1999.
39. J Davies. *Gen. Microbiol. 138*:1553–1559, 1992.
40. RD Fleischmann, MD Adams, O White, RA Clayton, EF Kirkness, AR Kerlavage, CJ Bult, JF Tomb, BA Dougherty, JM Merrick, K McKenney, G Sutton, W FitzHugh, C Fields, JD Gocayne, J Scott, R Shirley, L-I Liu, A Glodek, JM Kelley, JF Weidman, CA Phillips, T Spriggs, E Hedblom, MD Cotton, TR Utterback, MC Hanna, DT Nguyen, DM Saudek, RC Brandon, LD Fine, JL Fritchman, JL Fuhrmann, NSM Geoghagen, CL Gnehm, LA McDonald, KV Small, CM Fraser, HO Smith, JC Venter. *Science 269*:496–512, 1995.
41. FR Blattner, G Plunkett 3rd, CA Bloch, NT Perna, V Burland, M Riley, J Collado-Vides, JD Glasner, CK Rode, GF Mayhew, J Gregor, NW Davis, HA Kirkpatrick, MA Goeden, DJ Rose, B Mau, Y Shao. *Science 277*:1453–1474, 1997.
42. M Riley, MH Serres. *Ann. Rev. Microbiol. 54*:341–411, 2000.
43. CM Fraser, RD Fleischmann. *Electrophoresis 18*:1207–1216, 1997.
44. SE Celniker. *Curr. Opin. Genet. Dev. 10*:612–616, 2000.
45. C Heller. *Electrophoresis 22*:629–643, 2001.
45. H Loferer. *Mol. Med. Today 6*:470–474, 2000.
46. MB Eisen, PO Brown. *Methods Enzymol. 303*:179–205, 1999.
47. JL DeRisi, VR Iyer, PO Brown. *Science 278*:680–686, 1997.
48. M Chee, R Yang, E Hubbell, A Berno, XC Huang, D Stern, J Winkler, DJ Lockhart, MS Morris, SP Fodor. *Science 274*:610–614, 1996.
49. CA Harrington, C Rosenow, J Retief. *Curr. Opin. Microbiol. 3*:285–291, 2000.
50. A Pandey, M Mann. *Nature 405*:837–846, 2000.
51. MP Washburn, JR Yates. *Curr. Opin. Microbiol. 3*:292–297, 2000.
52. AL Barry. *Clin. Lab. Med. 9*:203–219, 1989.
53. M Handfield, RC Levesque. *FEMS Microbiol. Revs. 223*:69–91, 1999.
54. MB Schmid. *Curr. Opin. Chem. Biol. 2*:529–534, 1998.
55. MB Schmid, N Kapur, DR Isaacson, P Lindroos, C Sharpe. *Genetics 123*:625–633, 1989.
56. N Judson, JJ Mekalanos. *Trends Microbiol 8*:521–526, 2000.
57. IR Singh, RA Crowley, PO Brown. *Proc. Natl. Acad. Sci. USA 94*:1304–1309, 1997.
58. CA Hutchison, SN Peterson, SR Gill, RT Cline, O White, CM Fraser, HO Smith, JC Venter. *Science 286*:2165–2169, 1999.

59. V Smith, D Botstein, PO Brown. *Proc. Natl. Acad. Sci. USA 92*:6479–6483, 1995.
60. J Bender, N Kleckner. *Proc. Natl. Acad. Sci. USA 89*:7996–8000, 1992.
61. DJ Lampe, ME Churchill, HM Robertson. *EMBO J 15*:5470–5479, 1996.
62. BJ Akerley, EJ Rubin, A Camilli, DJ Lampe, HM Robertson, JJ Mekalanos. *Proc. Natl. Acad. Sci. USA 95*:8927–8932, 1998.
63. N Judson, JJ Mekalanos. *Nature Biotech 18*:740–745, 2000.
64. I Miller, D Maskell, C Hormaeche, K Johnson, D Pickard, G Dougan. *Infect. Immun. 57*:2758–2763, 1989.
65. WS Reznikoff. *Annu. Rev. Genet. 6*:133–156, 1972.
66. A Newton, JR Beckwith, D Zipser, S Brenner. *J. Mol. Biol. 14*:290–295, 1965.
67. C Sharetzsky, TD Edlind, JJ LiPuma, TL Stull. *J. Bacteriol. 173*:1561–1564, 1991.
68. BA Dougherty, HO Smith. *Microbiology 145*:401–409, 1999.
69. Y Ji, B Zhang, SF Van Horn, P Warren, G Woodnutt, MKR Burnham, M Rosenberg. *Science 293*:2266–2269, 2001.
70. RB Sykes, DP Bonner. *Am. J. Med. 78(2A)*:2–10, 1985.
71. CT Walsh, SL Fisher, IS Park, M Prahalad, Z Wu. *Chem. Biol. 3*:21–28, 1996.
72. CE Barry, III, RA Slayden, AE Simpson, RE Lee. *Biochem. Pharmacol. 59*: 221–231, 2000.
73. RE Bruccoleri, TJ Dougherty, DB Davison. *Nucleic Acids Res. 26*:4482–4486, 1998.
74. RL Tatusov, DA Natale, IV Garkavtsev, TA Tatusova, UT Shankavarum, BS Rao, B Kiryutin, MY Galperin, ND Fedorova, EV Koonin. *Nucleic Acids Res. 29*:22–28, 2001.
75. G Perriere, L Duret, M Gouy. *Genome Res. 10*:379–385, 2000.
76. F Chetouani, P Glaser, F Kunst. *Microbiology 147*:2643–2649, 2001.
77. HW Mewes, D Frishman, C Gruber, B Geier, D Haase, A Kaps, K Lemke, G Mannhaupt, F Pfeiffer, C Schuller, S Stocker, B Weil. *Nucleic Acids Res. 28*:37–40, 2000.
78. A Krause, J Stoye, M Vingron. *Nucleic Acids Res. 28*:270–272, 2000.
79. G Yona, N Linial, M Linial. *Nucleic Acids Res. 28*:49–55, 2000.
80. C Goffin, J-M Ghuysen. *Microbiol. Mol. Biol. Rev. 62*:1079–1093, 1998.
81. MKB Berlyn. *Microbiol Mol. Biol. Rev. 62*:814–984, 1998.
82. CM Hamilton, M Aldea, BK Washburn, P Babitzke, SR Kushner. *J Bacteriol. 171*: 4617–4622, 1989.
83. AJ Link, D Phillips, GM Church. *J. Bacteriol. 179*:6228–6237, 1997.
84. F Arigoni, F Talabot, M Peitsch, MD Edgerton, E Meldrum, E Allet, R Fish, T Jamotte, M-L Curchod, H Loferer. *Nature Biotechnol 16*:851–856, 1998.
85. C Freiborg, B Wieland, F Spaltmann, K Ehlert, H Brotz, H Labischinski. *J. Mol. Biotechnol 3*:483–489, 2001.
86. MG Lorenz, W Wackernagel. *Microbiol. Rev. 58*:563–602, 1994.
87. SC Winans, SJ Elledge, JH Krueger, GC Walker. *J. Bacteriol. 161*:1219–1221, 1985.
88. SK Amundsen, AF Taylor, AM Chaudhury, GR Smith. *Proc. Natl. Acad. Sci. USA 83*:5558–5562.
89. KC Murphy. *J Bacteriol. 180*:2063–2071, 1998.

90. KA Datsenko, BL Wanner. *Proc. Natl. Acad. Sci. USA 97*:6640–6645, 2000.
91. D Yu, HM Ellis, E-C Lee, NA Jenkins, NG Copeland, DL Court. *Proc. Natl. Acad. Sci. USA 97*:5978–5983, 2000.
92. V Vagner, E Dervyn, S Dusko. *Ehrlich. Microbiology 144*:3097–3104, 1998.
93. A Campbell. In: *The Bacteriophage Lambda*. Cold Spring Harbor, NY: Cold Spring Harbor Laboratory Press, 1971, pp 13–44.
94. LS Haverstein, G Coomaraswamy, DA Morrison. *Proc. Natl. Acad. Sci. USA 92*: 11140–11144, 1995.
95. JP Claverys, A Dintilhac, E Pestova, B Martin, D Morrison. *Gene 164*:123–128, 1995.
96. M Stieger, B Wohlgensinger, M Kamber, R Lutz, W Keck. *Gene 226*:243–251, 1999.
97. JA Thanassi, SL Hartman-Neumann, TJ Dougherty, BA Dougherty, MJ Pucci. *Nucleic Acids Res. 30*:3152–3162, 2002.
98. M Xia, RD Lunsford, D McDevitt, S Iordanescu. *Plasmid 42*:144–149, 1999.
99. L Zhang, F Fan, LM Palmer, MA Lonetto, C Petit, LL Voelker, A St. John, B Bankovsky, M Rosenberg, D McDevitt. *Gene 255*:297–305, 2000.
100. M Jana, T-T Luong, H Komatsuzawa, M Shigeta, CY Lee. *Plasmid 44*:100–104, 2000.
101. F Fan, RD Lunsford, D Sylvester, J Fan, H Celesnik, S Iordanescu, M Rosenberg, D McDevitt. *Plasmid 46*:71–75, 2001.
102. RM Horton, SN Ho, JK Pullen, HD Hunt, Z Cai, LR Pease. *Methods Enzymol. 217*:270–279, 1993.
103. SN Ho, HD Hunt, RM Horton, JK Pullen, LR Pease. *Gene 77*:51–59, 1989.
104. J Rosamond, A Allsop. *Science 287*:1973–1976, 2000.
105. CA Harrington, C Rosenow, J Retief. *Curr. Opin. Microbiol. 3*:285–291, 2000.
106. H Gmuender, K Kuratli, K Di Padova, CP Gray, W Keck, S Evers. *Genome Res. 11*:28–42, 2001.
107. RA VanBogelen, KD Greis, RM Blumenthal, TH Tani, RG Matthews. *Trends Microbiol. 7*:320–328.
108. T Gaasterland, CW Swenson. *Trends Genet. 12*:76–78, 1996.
109. PD Karp, M Krummenacker, S Paley, J Wagg. *Trends Biotechnol. 17*:275–281, 1999.
110. A Bateman, E Birney, R Durbin, SR Eddy, RD Finn, EL Sonnhammer. *Nucleic Acids Res. 27*:260–262, 1999.
111. DB Searls. *Annu. Rev. Genom. Hum. Genet. 1*:251–279, 2000.
112. KD Young. *Biochimie 83*:99–102, 2001.
113. DC Hooper. *Clin. Infect. Dis. 31(suppl 2)*:S24–28, 2000.
114. DJ Jin, CA Gross. *J. Mol. Biol. 202*:45–58, 1988.
115. A Telenti, P Imboden, F Marchesi, D Lowrie, S Cole, MJ Colston, L Matter, K Schopfer, T Bodmer. *Lancet 341*:647–650, 1993.
116. I Chopra. *J Antimicrob. Chemother. 26*:607–609, 1990.
117. RN Jones. *Clin. Infect. Dis. 32(suppl 2)*:81–167, 2001.

7

Pathogenesis Genes as Novel Targets

Andrea Marra
Pfizer Global Research and Development, Groton, Connecticut, U.S.A.

The current crisis in medicine represented by antibacterial drug resistance is poised to send us back to the preantibiotic era if novel, potent, and safe drugs to treat infections are not developed soon. One approach that has been discussed as having some promise for success is that of identifying bacterial functions with specific roles in human infection and developing drugs to inhibit them. This strategy would represent a significant shift in antibacterial drug development, which historically relied on either whole-cell screens or targeted assays aimed at inhibiting a small number of targets involved in essential bacterial functions. The most successful and widely used antibiotics interfere with either bacterial cell wall synthesis, DNA replication, or protein synthesis; other drugs target the cell membrane, RNA polymerase, or a metabolic pathway. Drugs against proven targets such as these have been exploited to yield third- and even fourth-generation antibiotics in some classes, often with limited utility due to resistance issues. The need to develop novel drugs is undisputed; however, there does not appear to be a novel target class or pathway that has the appeal or promise of any of the proven targets. More widespread use of animal infection models and clever strategies to understand and identify virulence determinants begs our consideration of such functions as targets for antibacterial drug discovery.

There are several advantages to such an approach, making pathogenesis genes attractive targets: [1] virulence genes are likely to be specific to bacteria, with little chance of having close homologs in the human genome [1];[2] drugs targeting pathogenesis determinants will not necessarily interfere with bacterial growth, thereby perhaps removing the selective pressure for survival and subsequent resistance problems that plague traditional antibiotics [2];[3] antivirulence drugs may be used in conjunction with conventional therapies with possible synergistic effects or to "customize" more broad-spectrum antibiotics [1,3]; and [4] pathogenesis targets are likely to be novel since they may not be the same functions as those required for laboratory growth and thus are unlikely to already have drugs developed against them or be cross-resistant with current drugs.

There are however, several significant drawbacks to using these types of targets as the basis for antibacterial therapy. The major obstacle involves the difficulty of setting up screens to inhibit them *in vitro*; traditional minimal inhibitory concentration (MIC) screening would be ineffective in this scenario. The challenge will be to formulate assays in the absence of knowledge of function, *in vitro* activity, or phenotypic effects. In addition, it is unclear whether inhibition of these targets *in vivo* will have a significant enough effect to clear the infection. Given the distinctive nature of infections caused by different pathogens even within the same host site, it may be that an agent developed against a virulence factor from one organism will have no effect on a different organism. In other words, such targets may not be broad spectrum enough to be useful. This chapter focuses on current strategies to identify pathogenesis targets and how this formidable challenge for drug discovery might be overcome.

HISTORICAL PERSPECTIVE

It has long been observed that passage of virulent bacteria on laboratory media resulted in a decreased ability of those organisms to cause infection upon reintroduction into animals. This observation emphasizes the notion that bacteria are compact, efficient factories, generating only those products that are required for their immediate growth and/or survival. Such conservation of energy at times can result in genetic alteration, rendering the bacteria completely attenuated for virulence while their laboratory growth is unaffected. Indeed, genetic manipulations of bacteria have also generated mutants that show little to no growth defect *in vitro* but that are unable to cause infection in an animal model. Numerous examples of this phenomenon exist; one obvious explanation is that requirements for growth *in vivo* are more stringent than those *in vitro* such that mutations that are silent in rich laboratory media can have significant effects *in vivo*. Likewise it is possible that a mutation can have a more pleiotropic effect *in vivo* due to the complex environment of the host. Investigators have exploited this phenomenon by using gene banks to complement defects and restore virulence, thus identifying the gene(s) responsible for that trait. Complementation of mutant strains helped to identify several genes required for full virulence in a number of organisms.

Cell Culture Models

Early studies of pathogenesis relied on cell culture to dissect the multiple aspects of virulence into its individual components. Cell lines and assays were developed to study bacterial adhesion; uptake/internalization/invasion; evasion of host defenses such as phagocytosis, complement binding, and phagosome–lysosome fusion; survival and growth within cells; and host cell killing in a variety of systems. These methods, combined with the power of bacterial genetics, enabled researchers to identify bacterial factors important for these functions. The use of cell culture models as surrogates for animal infection allowed the development of assays designed to study individual aspects of pathogenesis, and bacterial proteins required at the different stages were identified for a number of pathogens. It was not always the case, however, that bacterial factors identified via such screens were found to be required for *in vivo* pathogenesis [4]. Investigators were forced to deal with the fact that virulence is multifactorial and dynamic, and organisms have evolved redundant and/or compensatory functions to be able to adjust accordingly. By way of example, the invasin protein of *Yersinia pseudotuberculosis* was identified in cell culture assays as being critical for cell adhesion [5]. Invasin was the first demonstration that a single protein could, when expressed in noninvasive *Escherichia coli*, render that organism invasive. The expectation was that an invasin mutant would be severely attenuated *in vivo*. In fact, further studies showed that an invasin loss-of-function mutant was still virulent in murine infection models and that *Y. pseudotuberculosis* has evolved alternate pathways to infect its host [6,7]. It soon became clear that cell culture screens do not take into account the temporal aspect of pathogenesis, such that a protein involved at an early stage in infection may not be required at later time points or for survival in another host site [4].

Screens Using Animal Models of Infection

In order to fully understand the multiple facets of the infection process it was necessary to develop infection models in animals that more closely resemble human disease. Such models should ideally take into account the route of infection, the bacterial load, the tissue tropism observed, the course of the infection through the host, and the eventual outcome as compared to those in humans. There exists a wide range of animal infection models covering the major bacterial pathogens affecting humans, the use of which has facilitated antibacterial discovery but which have been used only recently to study pathogenesis at the molecular level. More widespread use of animal models has resulted in a pool of information on molecular pathogenesis with the potential to identify new bacterial targets.

The most significant limitations to animal models compared to cell culture models are due to those of cost, space, and ethics. For a genetics approach to be successful, a large collection of mutants must be screened for loss of a given phenotype, making an animal infection screen impractical. For these reasons, researchers have developed clever, high-throughput strategies to identify bacterial factors important for the establishment, maintenance, and outcome of infection, ostensibly to understand how the infection process occurs for a given organism, but also to potentially develop inhibitors of that process for use as antibacterial therapy.

Two general approaches have been taken: one involves a mutagenesis protocol and the other involves strategies for studying conditional (*in vivo*) gene expression. In the first scenario, signature-tagged mutagenesis (STM), a pool of uniquely tagged mutants is used to infect animals, and, via comparison of input and output populations, attenuated mutants may be identified. In the second scenario, either a library of promoter fusions is screened for expression via cell sorting (differential fluorescence induction, or DFI) or selection (*in vivo* expression technology, or IVET), or a global analysis of mRNA is performed following infection (microarrays or transcript profiling).

Comparisons of Some of the Different *in Vivo* Strategies

Beneficial features common to all these approaches include the feasibility of tapping into multiple host sites at different times following infection, the adaptability to high-throughput format, and the ability to work in a wide range of genetically tractable gram-negative and gram-positive bacteria. The advantages and cautions of the different strategies are compared more broadly in Table 1.

STM. This method relies on the power of a mutant selection to identify bacterial genes required during infection of an animal [8]. The perennial limitation to using animals to screen mutants has been that mutants must be screened individually; STM avoids this by employing pools of uniquely tagged mutants in an *in vivo* negative selection scheme to identify those mutants that are unable to cause infection. This enables the investigator to infect mice with up to 100 different mutants each. As a direct consequence of this, each infection is in effect a competition experiment, whereby small differences in virulence between mutant strains are amplified in the context of a pool of pathogenically wild-type clones. In addition, the infection model chosen can influence the outcome. In the case of a model in which the organisms must disseminate to deeper tissues, or traverse some barrier in order to access their preferred niche, it is likely that some clones in the diverse infection pool would be killed or otherwise lost prior to spreading to the site that is sampled. As a result, such clones may be scored as being attenuated and yet when screened individually are found to be virulent. Technical problems due to the complexity of the pools of tags have arisen, requiring researchers to either purify 96 unrelated tags or to infect with a lower number of tagged mutants. Last, because this is a mutant screen, all mutants involving *in vitro* essential genes would be lost prior to infection.

However, despite these caveats, STM has been applied to numerous organisms, with the result that known virulence determinants as well as proteins involved in every cellular function have been identified (Table 2). Most of these

TABLE 1 Comparison of Recent Approaches to Identify *in Vivo* Important Targets

Strategy	Advantages	Limitations
IVET and RIVET	Is a selection, so downstream screening is minimized Not necessarily limited in the number of clones that may be screened at once	Threshold level of induction must be reached to overcome selection Can be used to identify induced genes only not down-regulated genes Mutants must be generated in order to assess the role of the identified genes in virulence
STM	Can be applied to multiple models to identify functions important for one infection or many Is a selection	Is a mutant screen—*in vitro* essential genes will be missed False negatives due to clonal effects competitive indices complexity of tags Potential for clones to transcomplement each other *in vivo*
DFI	Can arbitrarily set thresholds so as not to exclude lower expressing promoters Does not depend on "off" condition Is not a selection Can identify both up- and down-regulated genes Not limited by the number of clones that may be screened at once	Difficulty of sorting bacterial cells harvested from mammalian tissues; Gfp stability Variability of *gfp* expression depending on growth phase of cells Mutants must be generated in order to assess the role of the identified genes in virulence
Microarrays (transcriptional profiling)	Global analysis Is not a selection Can study both host and pathogen gene expression during infection	Expensive to set up Vast amounts of data to analyze

TABLE 2 Summary of *in Vivo* Strategies Applied to Selected Bacterial Pathogens

Pathogen	STM	IVET	DFI
Actinobacillus pleuropneumoniae	✓(38)	✓(39)	
Brucella abortus	✓(40)		
Brucella melitensis	✓(41)		
Escherichia coli K1	✓✓(42,43)		✓(44)
Legionella pneumophila	✓(45)		
Listeria monocytogenes	✓(46)	✓(47)	
Mycobacterium tuberculosis	✓✓(48,49)		
Neisseria meningitidis	✓(50)		
Pasteurella multocida	✓(51)		
Proteus mirabilus	✓(52)		
Pseudomonas aeruginosa	✓(53)	✓(54)	
Pseudomonas putida		✓(55)	
Salmonella typhimurium	✓(56)	✓✓✓(11,57,58)	✓(17)
Staphylococcus aureus	✓✓✓(9,59,60)	✓(61)	✓(62)
Streptococcus gordonii		✓(63)	
Streptococcus pneumoniae	✓✓(64,65)		✓(66)
Vibrio cholerae	✓(67)	✓(68)	
Yersinia enterocolitica	✓(69)	✓(70)	
Yersinia pseudotuberculosis	✓(71)		

latter proteins had never before been implicated in virulence. Some of the more interesting findings came about when multiple infection sites were polled [9]; these sorts of experiments allow one to determine whether specific proteins are required to infect certain niches, whereas others are involved more broadly in survival, dissemination, and replication functions regardless of the particular host site.

IVET and RIVET. IVET [10,11] is a selection strategy based on complementation of a defined auxotrophy. The auxotrophy [either *thyA* [12] or *puraA* [10] have been used successfully] renders the organism avirulent, and the selectivity comes from a promoter-trap library cloned upstream of the complementing gene lacking its own promoter. When auxotrophic organisms carrying this library are introduced into an animal, the majority of the bacteria will be unable to survive; however, those clones carrying a fragment of DNA containing a promoter that is up-regulated in the host will express the complementing gene and thus survive and multiply. Several permutations of IVET have been reported, including those that rely on antibiotic resistance or recombinase [RIVET [13,14]] as a reporter, depending on the organism of interest. Each permutation is a subtle variation on the original theme, yet each can add value by overcoming some of the limitations of the original. For example, RIVET (recombinase-based *in vivo* expression technology) removes the selective pressure and allows the identification of transiently expressed genes. In a twist on a traditional application, a recent report [15] used RIVET in *Vibrio cholerae* to identify regulators of *toxT* and *ctxA* that function *in vivo* to direct virulence.

Yet another variation of this is *in vivo* induced antigen technology (IVIAT) [16], whereby sera from infected patients are pooled and used to screen an inducible expression library to identify clones carrying host-induced genes. Unlike classical IVET, IVIAT is not a selection. This method allows the screening of potentially thousands of clones in a high-throughput fashion following a human infection, one of the rare methods to date to do so.

DFI. DFI is a promoter-trap screen for genes induced under various *in vitro* or *in vivo* conditions [17,18]. The screen relies on a promoterless *gfp* (green fluorescent protein) gene cloned downstream of random chromosomal fragments from the organism of interest. If a specific DNA fragment contains a promoter active under a given growth condition, *gfp* will be expressed and such clones can be sorted on the basis of their fluorescence using flow cytometry. Unlike the IVET system, DFI is not a selection, and so a threshold level of expression is not required for cell viability at any stage of the screen. This allows investigators to isolate promoter clones with even modest inducing effects. In addition, this screen does not necessarily depend on a strict "off" reading in the noninducing condition to be successful. On the other hand, because it is not a selection, it is necessary to individually screen hundreds of sorted clones to identify particular clones of interest. In addition, the stability of Gfp makes it difficult to use DFI to determine whether promoters are only transiently expressed, and it has been observed that *gfp* expression can vary in some bacteria depending on the growth phase of the cells.

Genomics and Microarray Technology. The vast amount of information that will be obtained through genomics and microarray technology (expression profiling) promises to have an enormous impact on antimicrobial drug discovery. With nearly 40% of genes from sequenced organisms having no assigned function a wealth of novel targets is waiting to be identified. Microarrays can point toward functions for unknown genes; likewise, by bundling gene expression profiles we may gain understanding of potential regulons [19,20]. In addition, as we learn more about bacterial requirements for and during infection we can identify genes and pathways needed for pathogens in a given niche or host environment. One recent study compared gene expression profiles of *Pseudomonas aeruginosa* grown in broth culture with that grown in a biofilm [21]. It is known that many bacterial infections involve biofilm formation, and the assumption is that significant differential gene expression is required for the bacteria to convert from a planktonic state to a biofilm. However, the authors found that only about 1% of genes were differentially expressed, with half being induced and half repressed in the biofilm condition.

That microarray technology may be applied to human infections opens up a huge opportunity to study aspects of disease never possible before. Bacteria (or bacterial mRNA) can be recovered from infected human tissue and gene expression analyzed [22], organisms identified, and antibiotic susceptibilities determined. These studies are really an extension of IVET and DFI, but on a more global scale, potentially in a human host, and with the addition of gene annotation to glean maximal information. A potential drawback to wide use of this technology is the small numbers of bacteria that can routinely be recovered from infection sites [19]. However, as technology progresses and more sensitive tools are developed, this should no longer be a hindrance.

Other Approaches. One interesting approach makes use of the fact that a single pathogenic bacterium can cause disease in a wide range of genera, with the hope that identifying common virulence determinants will provide greater understanding of the infection pathway for this organism as well as provide surrogate models in which to study infection [23–26]. It has been shown that a clinical isolate of *Pseudomonas aeruginosa* (PA14) can infect mice in a burn model, cause lesions in an *Arabidopsis thaliana* leaf infiltration model, and kill *Caenorhabditis elegans* when it is used as a food source [24,26]. The authors demonstrated that several nonessential genes act as virulence determinants in all three systems. More recent work has shown that *C. elegans* is also susceptible to killing by gram-positive bacterial pathogens, implicating specific proteins as being important for this pathogenesis [27]. It is still too early to tell whether these types of surrogate infection models will have utility as screens for antibacterial therapies or indeed whether this approach can be applied to a more broad range of bacterial pathogens. Such studies can lead to the identification of highly conserved virulence mechanisms and, through the use of a genetic system such as *C. elegans*, a means for screening potential compounds for *in vivo* efficacy and drug resistance mechanisms. Though the *C. elegans* and *A. thaliana* models are far removed from human disease, their usefulness as *in vitro* assays for high-throughput screening of compounds may lie in the ability to observe bacterial killing or growth inhibition in the absence of an MIC screen and outside of an animal.

A number of groups are extending these studies by attempting to identify genes whose expression is induced during human infection. For obvious reasons, the only reagents available for such studies are patient sera and infected tissue samples. One group [28,29] described a surface polysaccharide [poly-*N*-succinyl-β-1-6 glucosamine, (PNSG)] of *Staphylococcus aureus* that is preferentially expressed during mouse and human infections. This antigen was identified in lung tissue samples taken from *S. aureus*-infected patients using specific immune rabbit serum. When the same antiserum was used to probe *S. aureus* isolates grown under *in vitro* conditions, antigen expression was found to be decreased, indicating specific expression of the polysaccharide in the host. Subsequent studies indicated that this antigen would be a good vaccine target, and indeed, immunization with PNSG protected mice from challenge with *S. aureus*. All *S. aureus* strains examined contained the genes involved in PNSG biosynthesis, *icaADBC*. A surprising finding in light of these results is that *S. epidermidis* [30] and *S. aureus* (A. Marra and S. Ho, unpublished observation) strains carrying deletions of the *ica* locus are still capable of causing infection. This would indicate that either the expression of this antigen is model specific or that an inhibitory antibody perhaps can block more than PNSG, thus leading to attenuation of virulence.

In Vivo Targets: Some Evidence for Success

As the era of genomics progresses and more information is gained on the genetic structure of a multitude of bacterial pathogens, it is likely that a target's attractiveness will be determined as much by its importance in the disease process as its spectrum. Several common themes in the pathogenic arsenal are emerging—namely two-component signal transduction systems (TCSTS's), quorum sensing systems, and type III secretion systems. These may not have direct sequence homologies among pathogens but could easily have structural and functional homologies that may be exploited and to which small molecule inhibitors may be directed. TCSTS's are of great interest due to their capacity to regulate bacterial gene expression [31,32] and the finding that several have been shown to be essential [33,34]. It is known that in a wide range of gram-positive and gram-negative pathogens, quorum sensing plays a role in virulence gene regulation by coordinating gene expression of a population of cells [35]. The AgrC/AgrA system of *S. aureus* and the LasI/LasR system of *P. aeruginosa* are two notable examples. In the case of type III secretion systems, used by bacteria to inject potentially toxic molecules into host cells, it is often the case that the secretory apparatuses are highly conserved, whereas the injected molecules vary in structure and function [36]. The recent identification of a type III-like function in a gram-positive pathogen may serve to broaden the spectrum of this highly appealing target [37].

Vaccines represent a sort of proof-of-concept that virulence factor inhibition may be a successful strategy against bacterial pathogens [3]. The majority of current vaccines against bacterial virulence factors are aimed at either preventing colonization or blocking the function of a toxin. Such vaccines have been highly effective in preventing incidence and spread of disease, thereby lending support to the idea of inhibiting virulence factor production or function to fight infection.

Information Obtained from *in Vivo* Strategies

The most comprehensive analyses to search for *in vivo* targets have been performed with STM, IVET, and DFI. These methods have been applied to a wide variety of gram-negative and gram-positive bacterial pathogens (Table 2), many using different animal infection models, with some overlap in results in terms of identified genes (Table 3). Using different approaches or the same approach in a different animal model is likely to lead to a greater understanding of the genetic requirements for virulence in a given organism. What is striking from these studies is that they do not point toward a particular gene, or class of genes, that would be an ideal target(s) for antibacterial therapy.

TABLE 3 Select *S. aureus* Genes Identified in Multiple *in Vivo* Screens and Attenuation of Mutants

Gene	STM/ attenuated?	RIVET/ attenuated?	DFI/ attenuated?
asd	→*/yes[a]		
Coenzyme A disulfide reductase	→/yes		§/yes
Copper-transporting ATPase *copA*		£/ND	§/no[b]
Glycerol ester hydrolase		£/ND	§/no
Glutamyl endopeptidase	→/yes		§/+/−
Lipase precursor	→/+/−		§/no
Asparaginyl tRNA synthetase	→/yes		§/+/−
lspA	→*/yes		
lysA	→*/yes		
odhB	→*/yes		
oppF	→*/yes		
pmsR	→*/+/−		
trpB	→*/yes		

→ Gene identification by STM as reported by (9).
* Gene identification by STM as reported by (59).
£ Gene identification by RIVET as reported by (61).
§ Gene identification by DFI as reported by (62).
[a] At least 10-fold attenuation in at least one animal model.
[b] No attenuation compared to wild-type.
+/−, different attenuation levels depending on animal model.
ND, not determined.

Table 3 summarizes some overlapping results following published analyses of *S. aureus* using STM, IVET, and DFI. A number of genes were identified either in more than one application of a given screen or in different screens entirely. No single category of genes stands out as being isolated a disproportionate number of times from these screens, which is most likely indicative of the somewhat subtle requirements imposed by each. The disparate genes identified in the different screens also reflect the strain and model differences utilized by the different groups. A gene must be mutated in order for its role in virulence to be assessed. As expected, because it is a mutant screen, most of the genes identified by STM are important for virulence, since mutants in those genes show attenuation. In contrast, the two other strategies assume that genes expressed at a high level *in vivo* are important for infection. At any rate, no gene was identified in all three screens, which would require that a gene is not only induced but is essential for virulence in all the models examined, which include systemic, kidney abscess, subcutaneous abscess, wound, and rabbit endocarditis.

In combination these approaches have yielded targets and methods for understanding pathogenesis on a much larger scale than was possible even as late as the early 1990s. There has been a significant overlap in identified genes but also it is clear that these screens have the potential to identify genes that are model and/or method specific. Perhaps the complementarity achieved by these approaches will impact the future of the field and enable researchers to look at pathogenesis with a fresh eye.

Conclusions and Forecast

Since the early 1990s the rise of innovative, creative approaches to molecular pathogenesis, with great potential for identifying novel targets for antibacterial therapy. A number of intriguing targets have been isolated by these *in vivo* strategies; however, none has yet led to a successful drug candidate. That is not to say that this is an impossible task or an unfeasible approach. It is far too early at this stage to be able to predict the success of these strategies, but no other current or past approach is showing more promise.

With the molecular tools and infection models that have recently been developed, we are poised to reap the benefits. Perhaps more specific models would help move this area forward—allowing us to consider bacterial adaptations *in vivo*, as well as niche-specific, host-specific, and strain-specific responses. A genetic switch that would allow one to turn gene expression on or off *in vivo* would be a tool of enormous value. Surrogate models such as those discussed above may be helpful in target identification and high-throughput screening of antibacterial compounds.

Probably the main reason for avoiding this approach is the inability to generate MIC's and demonstrate *in vitro* activity of compounds directed against virulence targets. In some cases it may be possible to identify *in vitro* conditions under which inhibition of a virulence target would be lethal. It may be that the traditional screening paradigms need to be adapted in order to accommodate this new range of targets. This issue should be given serious consideration in the near future.

Perhaps the best bet for compounds against these targets would be in adjunct therapy in order to add specificity to current drugs. Virulence targets can have an impact on antibacterial drug discovery and all avenues should be explored. This field is still relatively new and further research will be needed to validate both the strategies and the targets they have generated, but the more significant lesson here is the necessity of developing tools for studying pathogenesis in the human host or appropriate surrogates to do so. As we gain a better understanding of molecular pathogenesis, we can only be in a better position to combat the pathogens.

REFERENCES

1. LE Alksne, SJ Projan. *Curr. Opin. Biotechnol. 11*:625–636, 2000.
2. JC Lee. *Trends Microbiol. 6*:461–463, 1998.
3. R Goldschmidt, M Macielag, D Hlasta, J Barrett. *Curr. Pharm. Design 3*:125–142, 1997.
4. V Miller. *Trends Microbiol. 3*:69–71, 1995.
5. R Isberg, D Voorhis, S Falkow. *Cell 50*:769–778, 1987.
6. J Pepe, M Wachtel, E Wagar, V Miller. *Infect. Immun. 63*:4837–4848, 1995.
7. A Marra, R Isberg. *Infect. Immun. 65*:3412–3421, 1997.
8. M Hensel, JE Shea, C Gleeson, MD Jones, E Dalton, DW Holden. *Science 269*: 400–403, 1995.
9. SN Coulter, WR Schwan, EY Ng, MH Langhorne, HD Ritchie, S Westbrock-Wadman, WO Hufnagle, KR Folger, AS Bayer, CK Stover. *Mol. Microbiol. 30*:393–404, 1998.
10. MJ Mahan, JM Slauch, JJ Mekalanos. *Science 259*:686–688, 1993.
11. MJ Mahan, JW Tobias, JM Slauch, PC Hanna, RJ Collier. *Proc. Nat. Acad. Sci. USA 92*:669–673, 1995.
12. S Rankin, R Isberg. *Infect. Agents Dis. 2*:269–271, 1993.
13. A Camilli, DT Beattie, JJ Mekalanos. *Proc. Nat. Acad. Sci. USA 91*:2634–2638, 1994.
14. JM Slauch, A Camilli. *Methods Enzymol. 26*:73–96, 2000.
15. S Lee, S Butler, A Camilli. *Proc. Nat. Acad. Sci. USA 98*:6889–6894, 2001.
16. M Handfield, LJ Brady, A Progulske-Fox, JD Hillman. *Trends Microbiol. 8*: 336–339, 2000.
17. RH Valdivia, S Falkow. *Mol. Microbiol. 22*:367–378, 1996.
18. RH Valdivia, S Falkow. *Science 277*:2007–2011, 1997.
19. CA Cummings, DA Relman. *Emerg. Infect. Dis. 6*:513–525, 2000.
20. M Kato-Maeda, Q Gao, PM Small. *Cell Microbiol. 3*:713–719, 2001.

21. M Whiteley, MG Bangera, RE Bumgarner, MR Parsek, GM Teitzel, S Lory, E Greenberg. *Nature 413*:860–864, 2001.
22. R Rappuoli. *Proc. Nat. Acad. Sci. USA 97*:13467–13469, 2000.
23. LG Rahme, EJ Stevens, SF Wolfort, J Shao, RG Tompkins, FM Ausubel. *Science 268*:1899–1902, 1995.
24. LG Rahme, FM Ausubel, H Cao, E Drenkard, BC Goumnerov, GW Lau, S Mahajan-Miklos, J Plotnikova, M-W Tan, J Tsongalis, CL Walendziewicz, RG Tompkins. *Proc. Nat. Acad. Sci. USA 97*: 8815–8821, 2000.
25. S Mahajan-Miklos, M-W Tan, LG Rahme, FM Ausubel. *Cell 96*:47–56, 1999.
26. M-W Tan, S Mahajan-Miklos, FM Ausubel. *Proc. Nat. Acad. Sci. USA 96*:715–720, 1999.
27. DA Garsin, CD Sifri, E Mylonakis, X Qin, KV Singh, BE Murray, SB Calderwood, FM Ausubel. *Proc. Nat. Acad. Sci. USA 98*:10892–10897, 2001.
28. D McKenney, J Hubner, E Muller, Y Wang, DA Goldmann, GB Pier. *Infect. Immun. 66*:4711–4720, 1998.
29. D McKenney, KL Pouliot, Y Wang, V Murthy, M Ulrich, G Doring, JC Lee, DA Goldmann, GB Pier. *Science 284*:1523–1527, 1999.
30. H Shiro, G Meluleni, A Groll, E Muller, TD Tosteson, DA Goldmann, GB Pier. *Circulation 92*:2715–2722, 1995.
31. NG Wallis. *Curr. Opin. Anti-Infect. Invest. Drugs 1*:428–434, 1999.
32. M Macielag, R Goldschmidt. *Expert Opin. Invest. Drugs 9*:2351–2369, 2000.
33. R Lange, C Wagner, A de Saizieu, N Flint, J Molnos, M Stieger, P Caspers, M Kamber, W Keck, KE Amrein. *Gene 237*:223–234, 1999.
34. JP Throup, KK Koretke, AP Bryant, KA Ingraham, AF Chalker, Y Ge, A Marra, NG Wallis, JR Brown, DJ Holmes, M Rosenberg, MKR Burnham. *Mol. Microbiol. 35*:566–576, 2000.
35. M Miller, B Bassler. *Ann. Rev. Microbiol. 55*:165–199, 2001.
36. C Hueck. *Microbiol. Mol. Biol. Rev. 62*:379–433, 1998.
37. JC Madden, N Ruiz, M Caparon. *Cell 104*:143–152, 2001.
38. T Fuller, S Martin, J Teel, G Alaniz, M Kennedy, D Lowery. *Microb. Pathogenesis 29*:39–51, 2000.
39. T Fuller, R Shea, B Thacker, M Mulks. *Microb. Pathogenesis 27*:311–327, 1999.
40. P Hong, R Tsolis, T Ficht. *Infect. Immun. 68*:4102–4107, 2000.
41. P Lestrate, R Delrye, I Danese, C Didembourg, B Taminiau, P Mertens, X DeBolle, A Tibor, C Tang, J Letesson. *Mol. Microbiol. 38*:543–551, 2000.
42. J Badger, C Wass, S Weissman, K Kim. *Infect. Immun. 68*:5056–5061, 2000b.
43. J Martindale, D Stroud, E Moxon, C Tang. *Mol. Microbiol. 37*:1293–1305.
44. J Badger, C Wass, K Kim. *Mol. Microbiol. 36*:174–182, 2000a.
45. P Edelstein, M Edelstein, F Higa, S Falkow. *Proc. Nat. Acad. Sci. 96*:8190–8195, 1999.
46. N Autret, I Dubail, P Trieu-Cuot, P Berche, A Charbit. *Infect. Immun. 69*:2054–2065.
47. C Gahan, C Hill. *Mol. Microbiol. 36*:498–507, 2000.
48. L Camacho, D Ensergueix, E Perez, B Gicquel, C Guilhot. *Mol. Microbiol. 34*: 257–267, 1999.
49. J Cox, B Chen, M McNeil, WJ Jacobs. *Nature 402*:79–83, 1999.

50. H Claus, M Frosch, U Vogel. *Mol. Gen. Genet. 259*:363–371, 1998.
51. T Fuller, M Kennedy, D Lowery. *Microb. Pathogenesis 29*:25–38, 2000a.
52. H Zhao, X Li, D Johnson, H Mobley. *Microbiology 145*:185–195, 1999.
53. D Lehoux, F Sanschagrin, R Levesque. *Biotechniques 26*:473–478, 1999.
54. J Wang, A Mushegian, S Lory, S Jin. *Proc. Nat. Acad. Sci. 93*:10434–10439, 1996.
55. S Lee, D Cooksey. *Appl. Environ. Microbiol. 66*:2764–2772.
56. JE Shea, M Hensel, C Gleeson, DW Holden. *Proc. Natl. Acad. Sci. 93*:2593–2597, 1996.
57. D Heithoff, D Conner, P Hanna, S Julio, U Hentschel, M Mahan. *Proc. Nat. Acad. Sci. USA 94*:934–939.
58. A Janakiraman J Slauch, *Mol. Microbiol. 35*:1146–1155, 2000.
59. JM Mei, F Nourbakhsh, C Ford, D Holden. *Mol. Microbiol 26*:399–407, 1997.
60. WR Schwan, SN Coulter, EYW Ng, MH Langhorne, HD Ritchie, LL Brody, S Westbrock-Wadman, AS Bayer, KR Folger, CK Stover. *Infect. Immun. 66*:567–572, 1998.
61. AM Lowe, DT Beattie, RL Deresiewicz. *Mol. Microbiol 27*:967–976, 1998.
62. WP Schneider, SK Ho, J Christine, M Yao, A Marra, AE Hromockyj. *Infect. Immun. 70*:1326–1333.
63. A Kiliç, M Herzberg, M Meyer, Z Zhao, L Tao. *Plasmid 42*:67–72, 1999.
64. GW Lau, S Haataja, M Lonetto, SE Kensit, A Marra, AP Bryant, D McDevitt, DA Morrison, DW Holden. *Mol. Microbiol 40*:555–571, 2001.
65. A Polissi, A Pontiggia, G Feger, M Altieri, H Mottl, L Ferreri, D Simon. *Infect. Immun. 66*:5620–5629, 1998.
66. A Marra, J Asundi, M Bartilson, S Lawson, F Fang, J Christine, C Wiesner, D Brigham, WP Schneider, AE Hromockyj. *Infect. Immun. 70*:1422–1433.
67. S Chiang, J Mekalanos. *Mol. Microbiol. 27*:797–805, 1998.
68. A Camilli, J Mekalanos. *Mol. Microbiol. 18*:671–683, 1995.
69. A Darwin, V Miller. affecting survival in an animal host using signature-tagged transposon mutagenesis. *Mol. Microbiol. 32*:51–62.
70. G Young, V Miller. *Mol. Microbiol. 25*:319–328.
71. J Mecsas, I Bilis, S Falkow. *Infect. Immun. 69*:2779–2787, 2001.

8

Application of Genomics to the Discovery of New Drugs Against Tuberculosis

Richard A. Slayden, Dean C. Crick,
Michael R. McNeil, and Patrick J. Brennan
Colorado State University, Fort Collins, Colorado, U.S.A.

INTRODUCTION

Approximately 30 years ago, the prevailing opinion was that the fight against infectious diseases was all but over. Optimistic predictions were made in the belief that a combination of improved public health, vaccination, and the current variety of antimicrobial agents would be sufficient to combat most, if not all, infectious diseases. Synthetic analogs of known antibiotics that were effective against a broad range of organisms had been developed, and thus there was little perceived need for the development of new drugs. In fact, until very recently, no new classes of antibiotics had been licensed for clinical use against bacterial infections (1). The search for new drugs was limited to well-known compound classes with insufficient variability to prevent the escalation of clinical resistance (2,3).

With the increasing incidence of infection with multiple-drug-resistant tuberculosis strains, the World Health Organization (WHO) declared a worldwide emergency and responded with the directly observed therapy, short-course (DOTS) program (4–6). Streptomycin, isoniazid, and pyrazinamide, which were developed in the 1940s, are still integral parts of the chemotherapeutic regimens used today. The most recent introduction of an antituberculosis agent into the modern treatment scheme of tuberculosis was rifampin in the 1960s. The current treatment recommendations for tuberculosis from the Centers for Disease Control and the WHO for pan-sensitive strains involves a 6-month regimen, beginning with a 2-month (intensive phase) treatment cycle of isoniazid, rifampin, and pyrazinamide, followed by a 4-month cycle of isoniazid and rifampin. Although protracted, this is considered short-course chemotherapy (7). Alternative agents such as ethambutol and streptomycin are included in the intensive phase if there is a possibility of drug resistance (8,9).

The postgenomic era of expression profiling affords insight into the global biological response induced under diseaselike conditions in addition to drug pressure. Through the identification of gene and operon responses, specific pathways can be identified that are unique to a particular disease (10). To study bacteria in a diseaselike state has always been a difficult undertaking to achieve. Recently studies have been performed looking at the global gene response to altered growth states using microarray. These studies have been performed under artificial growth conditions to control the number of variables. In each case, specific changes were made, such as oxygen (11) and nutrient limitation (12). Although much information has been obtained at the level of the gene and related metabolic pathway, there is still not a clear picture of the impact this information has on the biology of the bacterium and thus the pathology of the disease. However, when these genes and pathways are analyzed, a picture develops of how the cell alters related regulons in order to compensate for the change in growth environment. The identification of these regulon networks results in information on the metabolome of the bacillus. This knowledge can be used to establish gene essentiality, conditional expression requirements, and determine genes associated with virulence. Although this information is extensive for the simple design of high-throughput screens, it affords predictive value in terms of the targets affected and the circumventive measures the bacillus uses to avert the pressure of drugs under different growth conditions such as the intracellular environment (13).

To some, high-throughput screening (HTS) has come to mean screening thousands of compounds with a pass–fail criterion. Ironically, none of the current antitubercular drugs could have passed the stringent conditions that are in place today. Therefore, we now realize that HTS is not intended to screen-out candidates but, rather, define rapidly and systematically the properties of many compounds so that they can be prioritized and further characterized through relevant secondary screens. It is important to clearly define the criteria that will be used to judge the candidates at each step of the drug discovery process. The overall goal is to make HTS a rational process such that assumptions made in the design do not compromise the quality of the data and therefore adversely impact the conclusions drawn from that data.

THE *MYCOBACTERIUM TUBERCULOSIS* GENOME AND ASSESSING THE IMPACT OF CHANGES IN ENVIRONMENT

The sequencing of complete genomes has had a profound influence on drug discovery, with its main promise being the potential to understand disease processes at a genetic level and determine optimal targets for drug intervention (14). Thus far over 1100 complete genomes have been sequenced, including 98 from bacteria and 16 from archaea (National Center for Biotechnology Information web site). This list includes many important human pathogens, although the bias is toward organisms that affect individuals in the developed countries (15). The complete *M. tuberculosis* genome was added to the list in 1998 (16). It is obvious that this sequence contains information on all possible drug targets encoded therein.

Genomic information is now an integral part of the modern drug discovery process and can be used interpretatively to understand protein and pathway function, allowing rapid target identification (17). This understanding is then used to establish a potential pathway to support the target of interest and virtually validate it. In addition to interpretive analysis, the greatest benefit of genome sequence is that it affords predictions toward establishing the metabolism of the bacterium. Usually researchers study *M. tuberculosis* on an artificial medium. Although the transcriptional regulators and associated networks can be used to predictively develop assays, they may not be reflective of the metabolism occurring in the organism during an infection and disease. The availability of the entire sequence information provides insight into the global metabolic capabilities and tendencies, which can be used to guide inhibitor design in a manner that may bias development toward relevant aspects of disease (14).

This approach has been deemed "structural genomics," and the aim is to structurally characterize proteins and thus pathways from sequence information (16,18). There is considerable effort in the postgenomic era to validate protein predictions with experimental determination *via* crystallography and NMR spectroscopy and synthetic pathways *via* experimental enzymology. However, currently the functional annotation of proteins and thus metabolic roles is based primarily on similarities to homologs in the databases (19). The primary assumption is that proteins with homologous sequences catalyze the same reactions. Although this is generally true, gene sequence cannot always be a reliable way to determine protein activity (20). Frequently, homologous enzymes do not catalyze the same chemical reaction and the same chemical reaction is not catalyzed by homologous enzymes. Although homologs can maintain function, there is mounting evidence that divergent evolution is an active process; a progenitor gene is duplicated and its copy assumes a new function in response to selective pressure (21). Because the outcome of divergent evolution can be quite different, functional assignment on the basis of homology could be grossly misassigned.

Although gene and protein homology can provide information about the nature of the reaction that is catalyzed, homology alone will provide very little information about the specific identity of the reaction (21). Therefore, genomic enzymology requires a correlation between structure (sequence) and chemical reaction (enzymatic function); neither structure-only-based nor function-only-based analysis will be totally accurate. This could be an underlying factor contributing to the high percentage (~40% in the case of *M. tuberculosis*) of genes that have no known function (16).

However, regardless of the difficulties associated with predicting all of the diverse chemical reactions present in the life of bacteria, including *M. tuberculosis*, genomic information has already proven invaluable in addressing a host of important research questions. A primary advantage of the availability of sequence information is that it has effectively made classical "reverse genetics" highly abbreviated. With genomic information, genes can be rapidly identified and cloned to evaluate function, and this approach is proving particularly useful when coupled with classical biochemical knowledge. Below we provide examples of this principle in the context of target definition and drug discovery. Pathways that can be exploited as novel and potent antibacterial targets are being mined *in silico* and this information is then applied to actual validation experiments.

Identifying genes, predicting protein function and pathway elucidation from preexisting biochemical data, is proving useful in determining the role of hypothetical and unknown genes through linkage (22,23). This linkage may be established based solely on location in the genome and proximity to known genes or by known function established through homology or experimentation. Ultimately, the information that is acquired contributes to the elucidation of specific and associated pathways, which may provide complete metabolome information. This then can provide clues about essential genes involved with disease, virulence factors, and conditional responses. Experiments designed to elucidate specific metabolic responses are rapidly developing into potential options for drug screening.

In an attempt to understand the disease process at a genomic level, studies have been performed to explore the altered metabolism of *M. tuberculosis* grown under defined conditions thought to mimic their intracellular niche (11,12). Classical biochemical studies have demonstrated that *M. tuberculosis* can survive extended periods of nonreplicating, stationary phase states, induced by limited growth conditions such as reduced oxygen and nutrient deprivation (24–26). However, until recently, these studies have not addressed the bacterial response at the gene-protein level.

Using microarray and proteome analysis, under nutrient starvation growth conditions, differential gene and protein expression were investigated (12). Proteome analysis derived from bacteria under these conditions revealed alterations in several proteins. Among repressed proteins were the 45-kDa and MPT64 antigens. It is interesting that these very proteins are the ones primarily recognized by the immune response of the majority of tuberculosis patients and their contacts (12,27). This phenomenon of the repression of the expression of two important immunogens under nutrient starvation may be explained by the fact that these are secreted proteins, and thus the artificial nutrient starvation conditions may affect protein secretion. It is not at all clear whether such artificially induced nutrient starvation conditions reflect long-term persistence or reactivation of disease. Several protein spots were shown to increase in intensity under starvation conditions. Among these are two conserved hypothetical proteins that have no identifiable functional motifs or domains when an *in silico* sequence-based analysis was performed (12).

In a more sensitive, perhaps more relevant and global approach to analyze the adaptive response of *M. tuberculosis* to nutrient starvation, the transcriptional profile was monitored at multiple time points using DNA microarrays (12). Differential gene expression was detected for all of the major functional gene classes. Genes associated with energy metabolism such as *aceAb* were upregulated, which may reflect altered carbon sources, although other tvilarboxylic acid cycle (TCA) related genes did not show consistent alteration. In accordance with reduced respiration rates and metabolic shutdown, genes thought to encode proteins involved in a shift from aerobic to anaerobic metabolism had altered expression profiles. Genes that are categorized into other functional groups, such as lipid and polyketide synthesis, translation, and transcription, responded accordingly to the slowdown in bacterial growth rate and metabolism. Of interest was the altered expression response of several unknown regulatory genes, and this may have the greatest potential in understanding the way *M. tuberculosis* responds and postures itself to surviving long-term starvation. Although a number of these genes have homologs and have been studied outside mycobacteria, the full understanding of their role in metabolic adaptation is as-yet unknown.

Sherman and colleagues (11) have also used DNA microarray technology to explore the adaptive response of *M. tuberculosis*. In these studies, gene response to hypoxic growth conditions were investigated. The gene expression changes that were noted were indicative of broad adaptations to reduced metabolic activity. Among the repressed genes, 40–60% (range is due to genes of unknown function) encode proteins with roles in biosynthesis, cell division, and aerobic metabolism. Genes induced by hypoxia tell a subtler story. Genes classically assigned to play a roles in *M. tuberculosis* latency, such as genes encoding glyoxylate shunt enzymes (28,29) and alternative sigma-factors (30,31), were essentially unaltered. However, of notable interest was the hypoxic induction of a two-component regulatory operon encoding Rv3134c/Rv3133c/Rv3132c. Providing more evidence that this regulatory operon is involved in hypoxic adaptation is the fact that targeted disruption of the upstream gene Rv3134c, eliminated hypoxic regulation of *hspx*. This is consistant with previous studies which clearly indicate the role of *hspx* in stationary-phase growth and growth in macrophages (32,33).

It is clear that the availability of genomic information has fostered the ability to delve into the intricacies of the bacterial response to a number of environmental changes. However, the current challenge is how to delineate expression profiles to provide an understanding of the disease state. Thus far, microarray and protein profiling have provided evidence for induction of the stringent response and downregulation of aerobic respiration, translation, cell division, and lipid biosynthesis, in response to hypoxic and nutrient-limited growth conditions. Induced genes such as *acr*, alanine dehydrogenase, and fumerate reductase may facilitate survival under these conditions and therefore may represent relevant drug targets for persistent organisms. In addition, the numerous genes of unknown function induced under these conditions may perform specific mycobacterial functions associated with invasion, pathology, and latent disease and warrant further investigation in terms of their relevance as potential drug targets.

TARGET-BASED DRUG DISCOVERY IN THE POSTGENOMIC ERA

Targeted screening allows the researcher to take advantage of the power of bioinformatics, which can be used to select potential targets from known or previously unexplored biochemical pathways. Once the target is selected, molecular biology techniques are used to clone and express recombinant forms of the protein. The resulting highly purified protein (usually an enzyme or a receptor) is then the foundation for the development of a high-throughput screen. As noted in Table 1, this method facilitates rational drug design, since it is often possible to utilize the purified recombinant protein for three-dimensional structural studies, which, in turn, allow structure–activity relationship (SAR) analysis (34). With a computational representation of the preferred molecular interaction, drug candidates can be screened, *in silieo,* for their ability to interact within this protein, which streamlines library screening. The power of target-based drug discovery is that very quickly the structure and the putative activity of a protein can be revealed. Then the specific characteristics can be applied to either understanding the pathway of interest to pursue related steps as targets or to developing an HTS-style drug screen. Since target-based assays are completely synthetic situations in which all of the mechanism kinetics can be predetermined, the assay can inform the researcher very easily about the inhibitory properties of a compound or class of compounds. This information can then be applied to combinatorial chemistry approaches to optimize inhibitory activity, as is discussed later. An additional advantage of target-based methods is that they allow detection of inhibitors that are not able to cross permeability barriers. Although these compounds are obviously not useful as therapeutics because they are not bioavailable nor able to

TABLE 1 Characteristics of Target-Based and Genome-Based Whole Cell Screening

Screening method	Advantages	Disadvantages
Target based	Can detect poorly permeable compounds that may be suitable for optimization Facilitates rational drug design Highly reproducible Compatible with combinatorial chemistry approaches	Cannot address potential targets of unknown function Assay development is time consuming (may be difficult to optimize for high throughput assays) *In vitro* inhibitors need to be converted into drugs Targets must be validated *in vivo*
Genome based whole cell	Selection for compounds that permeate cells Can establish antimicrobial properties Highly reproducible Historically successful Independent of preconceived targets Can address potential targets of unknown function Compatible with combinatorial chemistry approaches	Insensitive to compounds that do not permeate cells No rational basis for compound optimization Identification of target is difficult May have mixed mechanisms of action Does not predict pharmacokinetic issues

penetrate the bacterium, they do provide useful information that can be applied to discover other novel targets that are to be used to optimize new classes of inhibitors to improve their druglike properties.

Target-based screening does have some drawbacks. The method presupposes that a robust assay amenable to HTS can be developed. Development of assays of this nature can be time consuming and requires a considerable level of expertise. For this reason, target-based screening cannot address potential targets that have no known function. Thus, ~40% of the *M. tuberculosis* genome, some of which may be potential targets, is unavailable for target-based drug discovery. In addition, the essentiality of the potential target must be established *in vivo*, and this is not a trivial consideration when dealing with *M. tuberculosis* (35). Finally, compounds that inhibit an assay *in vitro* may not have druglike properties or may not be able to cross the permeability barrier imposed by the complex mycobacterial cell wall (36); turning an enzyme inhibitor into a drug is a very expensive endeavor in terms of both resources and time.

This approach in the context of tuberculosis is exemplified by targeting the L-rhamnose biosynthetic pathway in *M. tuberculosis*. L-Rhamnose plays an essential structural role in the cell wall of *M. tuberculosis* (Figure 1) as a key sugar component of the disaccharide "linker unit" (P-GlcNAc(α1–3)Rha), which connects the peptidoglycan heteropolysaccharide to the remainder of the cell wall "core," the galactan-arabinan-mycolic acid complex. In addition to its structural role in the cell wall, rhamnose biosynthesis was also chosen because it does not exist in mammalian cells, and, hence, inhibitors of its biosynthesis should have a lesser chance of being toxic to humans.

dTDP-Rhamnose is the precursor of the rhamnose found in the linker unit, and it is known that four enzymatic steps are required for its biosynthesis. Accordingly, four potential enzymes (RmlA-D) were sought through homology via *in silico* analysis of the *M. tuberculosis* genome. This analysis revealed eight ORFs, which could encode for proteins that catalyze these four enzymatic steps. Only one copy of *rmlC* (Rv3465) and *rmlD* (Rv3266c) were identified (16), and, upon cloning, the hypothesized enzymatic activity for each protein was demonstrated (37,38). The situation for *rmlA* was ambiguous because two candidate genes showed possible homology. However, upon cloning and expression, it was shown that Rv0334 encoded an active RmlA, whereas Rv3264c encoded a similar protein, ManB, but not involved in rhamnose synthesis. The situation was not as clear for the identification of the RmlB encoding gene. Through a series of molecular strategies, Rv3464 was determined to encode *rmlB*. One candidate, Rv3634c, encodes a protein with homology to RmlB, but the fact that its N-terminal sequence is nearly identical to UDP-galactose epimerase (GalE) from *M. smegmatis* (39), indicated that this gene encodes the GalE protein. Two other possible candidates, Rv3784 and Rv3468c, showed homology to *rmlB*; however, the enzymatic activity of the recombinant protein products could not be established, suggesting that these genes do not encode an active RmlB. Nevertheless, we cannot rule out the possibility that the soluble recombinant proteins are merely inactive forms of RmlB. Therefore, we assumed that there is only one isoform of RmlB present in *M. tuberculosis*, but this has not yet been established unequivocally.

The genome organization of the dTDP-rhamnose synthesizing (40) genes in mycobacteria revealed no clues in terms of their catalytic role. In most other organisms, such as *Escherichia coli*, *rmlA-D* are on a single operon, often in the order *B, D, A, C* (41–44). However, from the results above and the genome sequence it is clear that in *M. tuberculosis* the four genes responsible for dTDP-rhamnose synthesis are in three different loci. Thus, *rmlA* (Rv0334) is separate from all of the other genes involved in rhamnose metabolism and appears to be the fourth gene in an operon where the functions of the proteins encoded for by

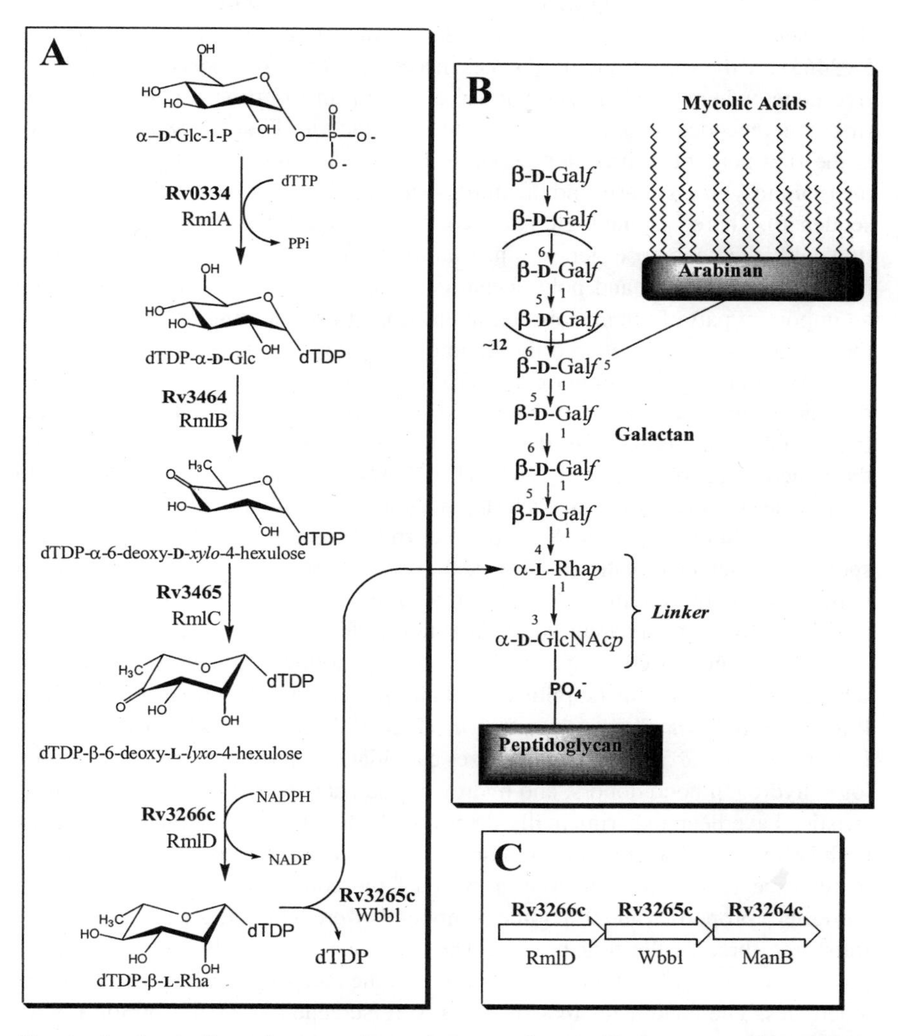

FIGURE 1 Anabolic pathway and proteins associated with rhamnose biosynthesis in *M. tuberculosis*. (A) The metabolic conversion of glucose-1-P to the rhamnose nucleotide, which is the direct precursor of cell wall rhamnose. (B) The macromolecular structure of the rhamnose containing "linker region" and arabinoglactan of the cell wall. (C) Operon of genes whose targets are being exploited as novel targets in *M. tuberculosis* drug discovery projects.

the other three genes are not known. The genes *rmlB* and *rmlC* (Rv3464 and Rv3465) are the second and third genes in a complex operon consisting of perhaps five genes where the last two genes are part of an insertion sequence and the first gene encodes for a protein of unknown function. Finally, *rmlD* (Rv3266c) is the first gene of a three-gene operon. In this case, the second gene is *wbbL*, the rhamnosyl transferase, and the third is *manB* (labeled as *rmlA-2* in the genome sequence). There is some logic in the coordinate expression of *manB* in conjunction with the rhamnose genes, since *manB* is needed for the synthesis of all mannosyl glycolipids and polysaccharides, which, like the rhamnosyl unit, are an important part of the mycobacterial envelope (45). The finding, however, that the *rml* genes are so scattered throughout the genome is surprising.

With the identification of each enzyme involved in the synthesis of dTDP-rhamnose, an assay was then developed. This assay took advantage of the fact that RmlD converts dTDP-6-deoxy-L-*lyxo*-4-hexulose to dTDP-rhamnose with the concomitant oxidation of NADPH to $NADP^+$, which makes it possible to couple the enzyme activity of each step in a facile microtiter plate assay amenable to screening inhibitors of the rhamnose biosynthetic pathway. Increased resolution specific to each enzymatic step was achieved by titrating the enzyme of interest within the assay system, thus altering required inhibitor concentrations. Upon analysis, it became apparent which enzymatic step was being inhibited.

This target-based assay, developed from genomic information, was used to screen 8000 compounds (supplied by Nanosyn, Tucson, AZ) in an HTS format. Prior to testing, *in silico* analysis of these candidates was performed based on Lipinski's "rule of 5," which categorizes candidates in terms of molecular weight, logP, hydrogen bond donors, and hydrogen bond acceptors (2,36). These characteristics have been experimentally determined and have been shown to correlate to the effective bioavailability of a drug from current orally available medicines. With respect to these four criteria, 89% of the compounds of the library had "druglike" properties and 9% of the compounds were within the acceptable range, satisfying three of the four criteria. The compounds were also selected for the presence of "druglike" functional groups and the lack of reactive groups such as aldehydes. The candidates that survived were screened at a concentration of 10 μM. Eighteen compounds, which showed activity initially, were rescreened in triplicate. To confirm the inhibitory activity of these 18 candidates, new samples were reexamined. However, upon rescreening, only 11 of the 18 compounds were inhibitory in the target-based assay. These 11 compounds were then assayed for their ability to specifically inhibit RmlB, RmlC, or RmlD, and all were shown to inhibit RmlC (dTDP-6-deoxy-D-xylo-4-hexalose epimerase). With specific inhibitory activity determined, these drug candidates will now proceed into whole cell-based assays to determine essentiality and inhibitory activity.

WHOLE-CELL-BASED SCREENING IN THE POSTGENOME CONTEXT

Classical antituberculosis drug discovery relied on whole-cell-based assays in which the primary readout was cell death. Although this approach has been responsible for the few anti-TB drugs in use today, the extended culturing times of *M. tuberculosis* (~4 weeks) made high-throughput screening impossible. However, since the genome sequence has become available, whole-cell-based screening has entered the era of HTS. A number of advances in our understanding of the biochemistry of the tubercle bacillus *via* genomic information have provided a scientific platform to design quick (~24 hr) bioluminescent-based HTS assays (Figure 2) (46–48). A primary breakthrough was the demonstration of the ease of expression profiling under stress conditions. The basic rationale in terms of modernized whole-cell-based screen development is that the selective inhibition of a metabolic pathways causes an accumulation of precursors and a concomitant depletion in products (2,49). These gross changes can be expected to selectively induce changes in the transcription of genes, thus altering the expression of the

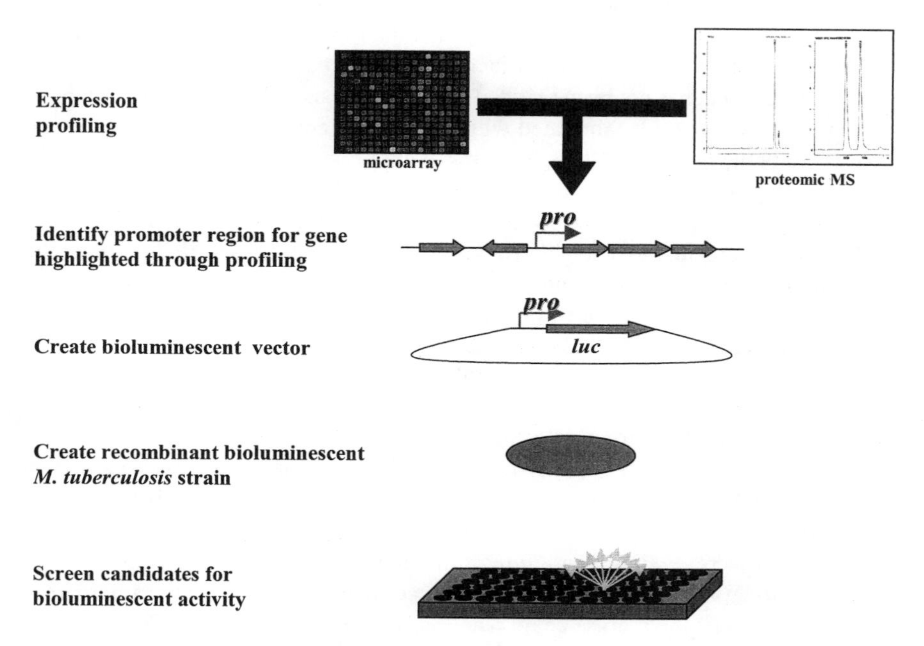

FIGURE 2 Design of high-throughput bioluminescence reporter strains used in *M. tuberculosis* drug discovery.

proteins that comprise the affected pathway, particularly before a more generalized stress response ensues. The resulting expression profile then serves as a "biological fingerprint" for that specific metabolic perturbation. While regulated genes themselves may or may not be targets for drug intervention, they, with the help of software algorithms linking genes with proteins and known functions, can point researchers toward known pathways. In addition, this information can be used to incriminate the affected pathway and classify the target for unknown inhibitors.

In terms of drug discovery and screening development, the expression of each gene in the genome or each protein in the proteome needs to be interrogated simultaneously in the presence and absence of the inhibitor (2). If an inhibitor of the pathway of interest is not available, then a manipulatable alteration needs to be introduced. One approach that we are currently employing is genetic engineering to construct temperature-sensitive strains of *M. tuberculosis* which can then be studied. This approach allows us to directly perturb the intended target protein, thereby introducing a "druglike" effect. Another strategy that has had limited success is the introduction of an inducible promoter to control the gene of the target protein. The limitation in the case of *M. tuberculosis* is attributable to the lack of an available good promoter that can be turned off when needed.

Currently, microarray technology is at the forefront of expression profiling in tuberculosis research because it is able to report on differential expression on a global scale that has not been realized by proteomic techniques (50,51). A typical *M. tuberculosis* array is fabricated with either amplicons or oligonucleotides and covers only the predicted ORFs. In terms of expression profiling of *M. tuberculosis* and drug discovery, the utility of transcriptional response profiling has been validated with the use of the *M. tuberculosis*-specific inhibitor, isoniazid (49). Although isoniazid has been thoroughly studied with arrays, it illustrates the important aspect of interpreting microarray data for screen development.

The mode of action of isoniazid, arguably, has been studied more than any other antimicrobial agent. Although the actual proteomic target has been debated, it is agreed that the affected macromolecules are the mycolic acids (3,52–56). When the transcriptional response of *M. tuberculosis* to isoniazid treatment was studied using DNA microarray technology (49), a subset of genes involved in fatty acid synthesis were differentially regulated. Many of these genes were assigned through biochemical studies and homology to fatty acid metabolism. The list includes genes of the FAS II operon Rv2243–2247, and associated genes *fbpC* (antigen 85c) (57), *fadE23*, and *fadE24*, in addition to isoniazid associated genes such as *efpA* (49) and *ahpC* (58–60). To confirm and validate the microarray results, real-time polymerase chain reaction (PCR) was performed. These experiments confirmed the expression profiles of all the genes tested. Our understanding of the molecular consequences of isoniazid on mycolate biosynthesis provides confidence that many of the responsive genes encode proteins that are relevant to the mode of action of the drug. Accordingly, microarray hybridization provides evidence useful in predicting the mode of action of a novel compound based on a physiologically derived interpretation of its expression response to that compound (49). Specific drug-responsive promoters identified by expression profiling serve as sensors of the intracellular conditions that are characteristic of the drug's activity which can then be exploited for the design of screens to identify novel compounds that exert similar effects on the bacteria.

Building on this approach, we have experimentally characterized the bacterial transcriptional response to over 20 inhibitors that affect several aspects of cell wall and macromolecular biosynthesis and cell division. Although many inhibitors clearly show similar transcriptional profiles, it is not always easy to rationalize the results from other inhibitors. Oftentimes, the highly induced genes are conserved hypothetical genes with no functional assignment. In such cases, it is hard to link a specific gene with the pathway affected by the inhibitor. Nonetheless, information is still generated that can be used as the basis for a mode of action specific high-throughput screen. We have been able to categorize compounds on the basis of common gene inductions. This not only indicates which drugs may affect similar biosynthetic pathways but also allows for the design of reporter strains of *M. tuberculosis* that can be used in an HTS format to rapidly evaluate the MIC of a molecule and characterize its possible mode of action. One has to remember, as previously discussed, that the premise of whole-cell, genome-based screens is that one is monitoring a unique phenotypic response, a "biological fingerprint" and not necessarily the targeted pathway.

INTEGRATION OF COMBINATORIAL CHEMISTRY INTO THE ANTITUBERCULOSIS DRUG DISCOVERY EFFORT

Modern combinatorial chemistry has already contributed to the modern drug discovery revolution (61). It is now possible to rapidly produce drug candidates with high levels of diversity and complexity, in marked contrast to rational drug design of just a decade ago in which chemists selected specific molecules for synthesis. This new chemical approach leads to compound libraries which have the ability, in theory, to sample all possible three-dimensional space

Barry and colleagues (62,63) have been instrumental in pioneering the use of combinatorial chemistry and expression profiling to discover new classes of inhibitors for *M. tuberculosis*. The beauty of this approach is that it capitalizes primarily on genomic sequence information which is readily available. The approach is centered on the idea that from genetic information, screens can be designed to report on the activity of drugs against a specific target, strictly from the reactive response of the bacterium. The advantage of this approach is that reporter strains (based on transcriptional response) can be designed very easily, and these are amenable to HTS automation. Progress is limited by classical structural methods such as crystallography (2).

To validate the whole-cell-based screening approach, Barry and colleagues used the known antimycobacterial drug ethambutol and interrogated the bacterial response to its inhibitory action (63). Although the target of ethambutol has been identified as arabinose polymerization in the cell wall, its specific protein target or targets are still unknown. In addition, ethambutol's potency could be improved; unlike other mycobacterial inhibitors, the toxicity profile is relatively low. These two factors allow for vast improvement of the drug's activity with a lesser chance of toxicity. With information of the transcriptional response to ethambutol treatment, a bioluminescent reporter strain was created and validated using ethambutol as a control. In addition to developing an ethambutol specific whole-cell-based HTS, a directed compound library was synthesized using combinatorial chemistry strategies. Ethambutol is a symmetrical diamine, and the directed library focused on exploring the activity of asymmetrical molecules around the diamine scaffold. The overall goal was to identify better inhibitors with the same mode of action as ethambutol using only genomic information. Numerous candidates were identified using HTS formatted MIC and bioluminescent assays that shared the same mode of action of ethambutol, but had better minimal inhibitory concentrations. The inhibitory activities of several of the candidates were confirmed using classical biochemical techniques and tested in the animal model of infection to evaluate efficacy. From analyzing the structure of all of the molecules in the library (compounds with and without inhibitory activity) in terms of their activity, it was possible to develop SAR information. Another round of molecular optimization can then be imitated biased by the SAR information to further improve the potency and the pharmacokinetic characteristics of the compounds. These first-generation molecules demonstrated the power of this approach in the development of new antimycobacterial drugs.

CONVERGENCE OF COMBINATORIAL LIBRARIES AND WHOLE-CELL- AND TARGET-BASED SCREENINGS

The convergence of chemistry and genomic information in combination with automation technologies has fostered a new approach to drug screening which utilizes both whole-cell-based and target-based approaches (64). In the scheme in Figure 3, a desired phenotypic change is sought by screening initially through high-throughput low-resolution whole-cell assays. These assays are designed based on unique patterns associated with specific pathway inhibition as determined by expression profiling. It is important to reiterate that the expression pattern may not implicate the actual proteomic target; it is simply readout of the bacterial response to the metabolic pressure. These screens have the ability to accommodate large libraries of diverse compounds and drastically reduce the

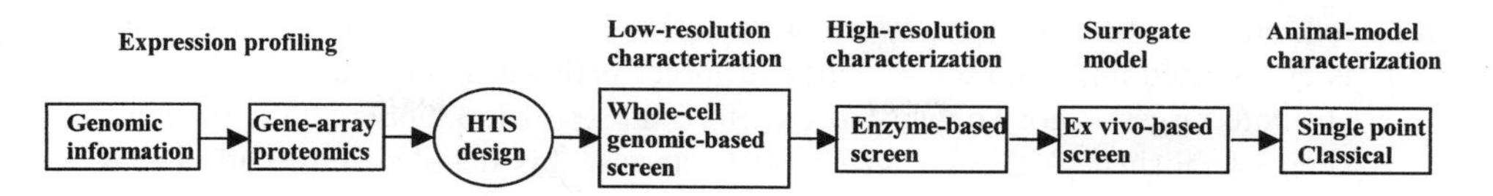

Figure 3 The modernized *M. tuberculosis* genome-based whole-cell screening approach.

number of candidates progressing through the discovery process. The "hits" identified from the whole-cell-based screens are then interrogated with higher resolution assays designed around specific enzymatic targets. These high-resolution assays provide high-quality data and establish inhibitory kinetics. Once the library has been narrowed to a small fraction of compounds with the desired effect, they are prioritized by surrogate animal models and finally evaluated for efficacy in an actual animal model. This drug screening process is designed so that the high-throughput low-resolution assays which are robust enough to handle large compound libraries are used before low-throughput high-resolution assays are used to confirm and bias the procession of candidates into model systems of infection.

As the number of potential drug candidates increases, so does the need for quick, efficient, and high-volume screens that generate data of high quality. The current testing management problem is that screening is not conducted on the same candidate a million times but on a million different candidates at the same time. The purpose of the drug discovery process is to find and explore significant differences between compounds. Therefore, the screening process has to be designed to categorize quickly the inhibitory characteristics of the drug candidates. The largest challenge in screening is not just the large member libraries, but the low hit rates associated with them. With the latest advances in high-throughput screening using multiwell plates, the initial testing volume can be much greater than 10,000 candidates a day. Modern primary screens are designed to accomplish two things: first, to establish a minimal inhibitory concentration for each candidate and, second, to provide clues as to the potential target. This can be accomplished by multiplexing the assays. The advantage of multiplexed whole-cell-based screens is that a couple of requirements are built into the assay. The drug candidate has the strict requirement to be able to subterfuge the mycobacterial cell wall and be metabolically stable within the cell. The primary benefit of whole-cell-based screens is to streamline the process and reduce the screening burden on target-based screens.

Potential drug candidates coming through the whole-cell-based assays can be prioritized and biased into the target-based assays based on their specific inhibitory profile. With the information on MIC, the ability to kill the bacterium is established and a reasonable idea of target pathway is known. However, the absolute target is not known and therefore the specific kinetic details have not been established. Although, in simplistic terms, this detailed information may not be necessary, it is required for predictive studies in terms of resistance mechanisms and potential microbial spectrum. There are a number of advantages in target-based screens, particularly if they are not responsible for the primary categorization of potential drug molecules. They provide better resolution in terms of target identification and they are an additional method of target validation.

Whole-cell-based assays and target-based assays are very compatible strategies in the quest to rapidly characterize potential molecules as drug candidates. They exploit both genomic information and expression profiling. A screening strategy that uses these two approaches in tandem has the ability to establish essentiality through MIC, elucidate targets, and establish kinetic information of the inhibited reaction. Despite all of these advantages, each approach has its disadvantages. Whole-cell-based screens can be insensitive; many active compounds have associated toxicity issues in general and possible mixed mechanisms of action. Target-based assays have problems associated with turning *in vitro* inhibitors into antibacterial drugs and genetic validation of essentiality. However, in combination, these two approaches compliment each another, and the disadvantages associated with each are essentially eliminated.

CORRELATING *IN VITRO* ACTIVITY WITH *IN VIVO* EFFICACY

The greatest need in the search for new antituberculosis drugs is a surrogate screening system for animal testing. Animal models of infection are used in the final stages of the drug development cycle to assess the antimicrobial efficacy in a model of human disease. Recent experience with library testing confirms historical evaluations of the extremely poor correlation between *in vitro* drug potency and *in vivo* drug efficacy (2,65,66). The reasons for this poor correlation may be due to (i) altered physiology of the bulk of the bacterial cells growing in macrophages; (ii) the simultaneous existence of several discrete populations of bacteria each with varying susceptibilities to the drugs being tested; and (iii) the extremely complex pharmacokinetics involved in delivering the drugs to the site of infection, namely the alveolar macrophages. A large number of compounds are being discovered and synthesized that show promise in *in vitro* assays against *M. tuberculosis*. In fact, many more *in vitro* lead molecules are being generated than can reasonably be assessed through traditional animal models of efficacy. Therefore, the development of reliable surrogate models for chemotherapy of human disease that has a moderately high-throughput format is an urgent matter (Table 2).

TABLE 2 Characteristics of Animal Models of Infection

Screen	Advantages	Disadvantages
Surrogate system	Establishes antimicrobial properties under bacterial metabolism conditions associated with disease Reproducible Prioritizes the procession of candidates Compatible with combinatorial chemistry approaches *In vitro* inhibitors need to be converted into drugs	No rational basis for compound optimization Targets must be validated *in vivo*. Identification of target is difficult Many antimicrobials are also toxic to eukaryotes Assay development is time consuming
Single-point animal model	Can detect poorly bioactive compounds Highly reproducible Prioritizes the procession of candidates Compatible with combinatorial chemistry approaches Fast and relatively easy	Cannot address properties of bioavailability Many potential inhibitors are also toxic to animals
Classical animal testing model	Detects for compounds that permeate cells Can establish antimicrobial efficacy in animal model Addresses properties of bioavailability	Not compatible with combinatorial chemistry approaches Time consuming Many potential inhibitors are also toxic to animals

Bioluminescent screening systems to address these questions have been attempted using the same basic approaches as the whole-cell assays. Currently, murine-derived macrophage cell culture is used as a possible testing surrogate (67,68). More recently, screens have been developed with bacteria derived from infectious animal tissue, such as the lung or spleen of a mouse (69). Since the bacteria have established an actual infection, they have also adopted the proper physiological state to cause disease and persist. The purpose is to establish this bacterial metabolic state so that drug candidates can be screened rapidly and easily in a system that mimics the bacterial state of disease. Although current approaches have utilized macrophage lines, it is best if this metabolic state could be induced in artificial medium. This does not seem to be possible at the moment, but, with genomic information and ongoing studies on *in vivo* metabolism of *M.*

tuberculosis, a manipulatable, quorum-sensing-based system may be possible in the future. It is important to remember that the primary quality from a surrogate animal system is that it establishes the bacterium in a metabolic state similar to that of true infection. This should not be confused with developing a system that tries to mimic the site of infection.

After the efficacy of a compound has been addressed in the surrogate model of infection, it is then tested in an animal model of infection. Due to the number of candidates that can proceed through a whole-cell-based discovery strategy to this step, low-cost, low-resolution, single-time-point animal experiments are useful. These animal tests are based on a single-time-point analysis, such as 40 days of infection. The animals begin therapy at 2 weeks postinfection, and the gross effects of drug efficacy are determined at the time of harvest. It is important to begin therapy before visible pathology is apparent. The bacterial burden, which reflects the efficacy of the drug candidate, can be qualitatively determined *via* gross pathology, and the quantities of bacteria can be determined through approaches such as bioluminescence and molecular tagging, if desired. This approach is intended, like the other screens, to bias the procession of candidates, sifting the most promising compound to the top of the pile. Candidates that show promising activity can be further analyzed *via* classical animal model experiments designed to determine the efficacy profiles of potential drugs over several weeks, consisting of multiple time points that provide high-quality statistical data.

THE NEXT GENERATION: INHIBITION OF THE MECHANISM OF AN ENZYME OR ENZYME MECHANISMS?

In terms of drug discovery, it is paramount to have functional knowledge derived from genomic sequence or experimental data to fully understand the inhibitory mechanism. With the genomes of so many organisms completed, it is becoming possible to globally compare organisms with similar disease niches and pathology. This genomic–metabolic linkage among pathogens is proving to be the aspect that is most promising to understanding the weakness of organisms in general. With this metabolic information, combinatorial libraries can be designed which are biased for unique activity specifically against reactions, not necessarily unique enzymes.

The next step to optimizing drug activity is to marry mechanistic and chemical information with the predicted function of the target protein based on genomic information and homology so that chemical libraries can be tailored to specific enzymatic chemistry. At the moment, whole-cell-based bioluminescent screening strategies have the potential to characterize large sets of candidates against a single enzyme within a pathway. This is slowly producing hits and resulting in the identification of novel inhibitors. However, they usually lead to a single target, and, with the treatment pressure by current drugs, they are rapidly losing their potency due to resistance. Thus, years of work and resources can be lost because a drug that has been designed against a single enzyme which undergoes a mutation event renders the drug useless.

Perhaps the next step is to use genomic sequence to inform drug design against specific enzymatic reactions that are represented numerously within the metabolism of an organism or multiple organisms. This would effectively make it impossible for an organism to acquire a single polymorphism that would provide resistance. Microorganisms develop resistance at a frequency as low as 10^5; if a number of targets were inhibited simultaneously, this mutation frequency would be multiplied based on the number of targets. Organisms have the ability to reach high numbers in an infection, which makes this frequency a very real number. The basic rationale is not to design inhibitors against a reaction catalyzed by a specific enzyme, but rather against a specific reaction catalyzed by protein analog super families.

Analysis of the mycobacterial genome sequence reveals that the organism has undergone massive gene duplication. This is particularly evident by the number of fatty acid genes, related polyketide syntheses (PKS), and glycosyl transferases that are responsible for cell wall construction and accessory molecules. Although all of these related enzymes catalyze similar reactions utilizing a variety of related substrates, they clearly produce very different products. This has essentially led to the characteristic heteropolymeric cell wall of the tubercle bacillus, which is made up of diverse polysaccharides and unique lipids.

Once a gene is duplicated, it is thought that it undergoes evolution to fulfill a new metabolic role. It has been proposed that metabolic pathways evolve backward. When the substrate of an enzyme in a pathway is limited, a new enzyme evolves to supply that substrate from an available precursor. Accordingly, evolution is constrained to retain binding specificity, because the original enzyme and the newly evolved enzyme must retain the ability to bind the same substrate/product. The specific mechanism of the reactions need not be conserved. Ultimately, the progenitor and new enzyme may catalyze the same reaction from slightly different substrates or catalyze a slightly different reaction from similar substrates, each case utilizing the same type of intermediate.

In *M. tuberculosis*, an example of the first idea may be the ketoacyl syntheses KasA and KasB and possibly the related domain of fatty acid synthase, Type I (FASI). Each one catalyzes the same reaction under the same conditions, the primary difference being substrate specificity; KasA specifically catalyzes the condensation of short-chain lipids and KasB condenses long-chain lipids. If molecules can be tailored to mimic the mechanistic state of the condensation in this case, then all three enzymes in addition to PKSs could be potential targets.

The family of enzymes responsible for the introduction of heterogeneity into mycolic acids may represent the second scenario. It has been suggested that these enzymes introduce different functionality into the hydrocarbon chain under a common transition state. In this case, the substrate and intermediate are the same; the primary difference in the reaction is how the reaction proceeds, therefore resulting in a different functional group. A synergistic strategy utilizing genomic information and evolutionary principles can be expected to provide a more efficient solution to drug development than the more traditional approach of studying a single enzyme at a time.

CONCLUSION

It was once thought that the "Holy Grail" in understanding the molecular basis of infectious disease was to know the total genetic sequence of the infectious organism with the promise of an understanding of the disease and pathology of each infectious agent at a genetic level. However, it is becoming obvious that deciphering and understanding all of the information to the point of assigning function is the largest hurdle. Having the genomic sequence of a bacterial organism allows one to mine the genome for information beyond predicting ORFs. The big push now is in correlating expression profiles with metabolic outcome. At the moment, there is a tremendous amount of raw sequence data and related information. Considering all of the preexisting biochemical information and differential expression data from DNA based arrays and proteomic studies, each piece of DNA sequence can now be coupled with this information.

Our aim with this chapter is not to rehash the themes of the past, but rather, with history, to spur all of us toward new directions and goals that push the limits of science, to nudge us to step out of the scientific comfort "pale" and explore new dimensions within the drug discovery arena in the postgenomic era. Therefore, the primary goal of genomics-based research in the *M. tuberculsois* drug discovery process is to identify new drug targets. Drugs emanating from these novel targets will provide great benefits in terms of treatment of MDR strains. This is obtainable because the efficiency of identifying targets from the genome has increased markedly, replacing the old process of identifying proteins, cloning them, and establishing a level of validation of their potential utility. This barrier has essentially been removed as a consequence of the completion of the *M. tuberculosis* genome and the development of *in silico* analysis tools. All proteins now become potential targets and can be prioritized on the basis of information indicating essentiality to cell survival, conserved range in pathogens, and presence in humans.

REFERENCES

1. DJ Knowles. *Trends Microbiol 5*:379–383, 1997.
2. CE Barry 3rd, RA Slayden, AE Sampson, RE Lee. *Biochem Pharmacol 59*:221–231, 2000.
3. RA Slayden, CE Barry 3rd. *Microbes Infect 6*:1–11, 2000.

4. World Health Organization. *TB—A Global Emergency: WHO Report on the TB Epidemic*. Geneva: WHO, 1994.
5. WHO/IUATLD. *Anti-Tuberculosis Drug Resistance in the World*. Geneva: WHO Global Tuberculosis Programme, 1997.
6. World Health Organization. *TB—Groups at Risk: WHO Report on the Tuberculosis Epidemic*. Geneva: World Health Organization, 1996.
7. MD Iseman. *Chemotherapy 45*:34–40, 1999.
8. MD Iseman. *Chemotherapy 45*:3–11, 1999.
9. JB Bass Jr, LS Farer, PC Hopewell, R O'Brien, RF Jacobs, F Ruben, DE Snider Jr, G Thornton. *Am J Respir Crit Care Med 149*:1359–1374, 1994.
10. JL DeRisi, VR Iyer, PO Brown. *Science 278*:680–686, 1997.
11. DR Sherman, M Voskuil, D Schnappinger, R Liao, MI Harrell, GK Schoolnik. *Proc Natl Acad Sci USA 98*:7534–7539, 2001.
12. JC Betts, PT Lukey, LC Robb, RA McAdam, K Duncan. *Mol Microbiol 43*:717–731, 2002.
13. MJ Marton, JL DeRisi, HA Bennett, VR Iyer, MR Meyer, CJ Roberts, R Stoughton, J Burchard, D Slade, H Dai, DE Bassett Jr, LH Hartwell, PO Brown, SH Friend. *Nat Med 4*:1293–1301, 1998.
14. J Rosamond, A Allsop. *Science 287*:1973–1976, 2000.
15. CM Tang, ER Moxon. *Annu Rev Genomics Hum Genet 2*:259–269, 2001.
16. ST Cole, R Brosch, J Parkhill, T Garnier, C Churcher, D Harris, SV Gordon, K Eiglmeier, S Gas, CE Barry 3rd, F Tekaia, K Badcock, D Basham, D Brown, T Chillingworth, R Connor, R Davies, K Devlin, T Feltwell, S Gentles, N Hamlin, S Holroyd, T Hornsby, K Jagels, BG Barrell, et al. *Nature 393*:537–544, 1998.
17. ST Cole. *FEBS Lett 452*:7–10, 1999.
18. RC Stevens, S Yokoyama, IA Wilson. *Science 294*:89–92, 2001.
19. D Baker, A Sali. *Science 294*:93–96, 2001.
20. JA Gerlt, PC Babbitt. *Genome Biol 1*:REVIEWS0005, 2000.
21. JA Gerlt, PC Babbitt. *Annu Rev Biochem 70*:209–246, 2001.
22. WJ Philipp, DC Schwartz, A Telenti, ST Cole. *Electrophoresis 19*:573–576, 1998.
23. ST Cole. *Curr Opin Microbiol 1*:567–571, 1998.
24. TP Primm, SJ Andersen, V Mizrahi, D Avarbock, H Rubin, CE Barry 3rd. *J Bacteriol 182*:4889–4898, 2000.
25. LG Wayne, KY Lin. *Infect Immun 37*:1042–1049, 1982.
26. LG Wayne. *Eur J Clin Microbiol Infect Dis 13*:908–914, 1994.
27. AT Kamath, CG Feng, M Macdonald, H Briscoe, WJ Britton. *Infect Immun 67*: 1702–1707, 1999.
28. JD McKinney, K Honer zu Bentrup, EJ Munoz-Elias, A Miczak, B Chen, WT Chan, D Swenson, JC Sacchettini, WR Jacobs Jr, DG Russell. *Nature 406*:735–738, 2000.
29. V Sharma, S Sharma, K Hoener zu Bentrup, JD McKinney, DG Russell, WR Jacobs Jr, JC Sacchettini. *Nat Struct Biol 7*:663–668, 2000.
30. J DeMaio, Y Zhang, C Ko, DB Young, WR Bishai. *Proc Natl Acad Sci USA 93*: 2790–2794, 1996.
31. J DeMaio, Y Zhang, C Ko, WR Bishai. *Tuber Lung Dis 78*:3–12, 1997.
32. Y Yuan, DD Crane, CE Barry 3rd. *J. Bacteriol 178*:4484–4492, 1996.

33. Y Yuan, DD Crane, RM Simpson, YQ Zhu, MJ Hickey, DR Sherman, CE Barry 3rd. *Proc Natl Acad Sci USA 95*:9578–9583, 1998.
34. DS Bailey, A Bondar, LM Furness. *Curr Opin Biotechnol 9*:595–601, 1998.
35. M Jackson, DC Crick, PJ Brennan. *J Biol Chem 275*:30092–30099, 2000.
36. CA Lipinski, F Lombardo, BW Dominy, PJ Feeney. *Adv Drug Delivery Rev 23*: 3–25, 1997.
37. TT Hoang, Y Ma, RJ Stern, MR McNeil, HP Schweizer. *Gene 237*:361–371, 1999.
38. RJ Stern, TY Lee, TJ Lee, W Yan, MS Scherman, VD Vissa, SK Kim, BL Wanner, MR McNeil. *Microbiology 145*:663–671, 1999.
39. A Weston, RJ Stern, RE Lee, PM Nassau, D Monsey, SL Martin, MS Scherman, GS Besra, K Duncan, MR McNeil. *Tuber Lung Dis 78*:123–131, 1997.
40. L Deng, K Mikusova, KG Robuck, M Scherman, PJ Brennan, MR McNeil. *Antimicrob Agents Chemother 39*:694–701, 1995.
41. SJ Lee, LK Romana, PR Reeves. *J Gen Microbiol 138*:1843–1855, 1992.
42. K Rajakumar, BH Jost, C Sasakawa, N Okada, M Yoshikawa, B Adler. *J Bacteriol 176*:2362–2373, 1994.
43. PR Reeves, M Hobbs, MA Valvano, M Skurnik, C Whitfield, D Coplin, N Kido, J Klena, D Maskell, CR Raetz, PD Rick. *Trends Microbiol 4*:495–503, 1996.
44. G Stevenson, K Andrianopoulos, M Hobbs, PR Reeves. *J Bacteriol 178*:4885–4893, 1996.
45. RE Lee, PJ Brennan, GS Besra. *Curr Top Microbiol Immunol 215*:1–27, 1996.
46. TM Arain, AE Resconi, DC Singh, CK Stover. *Antimicrob Agents Chemother 40*: 1542–1544, 1996.
47. MJ Hickey, TM Arain, RM Shawar, DJ Humble, MH Langhorne, JN Morgenroth, CK Stover. *Antimicrob Agents Chemother 40*:400–407, 1996.
48. TM Arain, AE Resconi, MJ Hickey, CK Stover. *Antimicrob Agents Chemother 40*: 1536–1541, 1996.
49. M Wilson, J DeRisi, H-K Kristensen, P Imboden, S Rane, PO Brown, GK Schoolnik. *Proc Natl Acad Sci USA 96*:12833–12838, 1999.
50. JD Amick, YV Brun. *Genome 2*:REVIEWS1020, 2001.
51. MT Laub, HH McAdams, T Feldblyum, CM Fraser, L Shapiro. *Science 290*: 2144–2148, 2000.
52. K Mdluli, RA Slayden, Y Zhu, S Ramaswamy, X Pan, D Mead, DD Crane, JM Musser, CE Barry 3rd. *Science 280*:1607–1610, 1998.
53. RA Slayden, RE Lee, CE Barry 3rd. *Mol Microbiol 38*:514–525, 2000.
54. A Banerjee, E Dubnau, A Quemard, V Balasubramanian, KS Um, T Wilson, D Collins, Gd Lisle, WR Jacobs Jr. *Science 263*:227–230, 1994.
55. LA Basso, R Zheng, JM Musser, WR Jacobs Jr, JS Blanchard. *J Infect Dis 178*: 769–775, 1998.
56. L Kremer, JD Douglas, AR Baulard, C Morehouse, MR Guy, D Alland, LG Dover, JH Lakey, WR Jacobs Jr, PJ Brennan, DE Minnikin, GS Besra. *J Biol Chem 275*: 16857–16864, 2000.
57. TR Garbe, NS Hibler, V Deretic. *Antimicrob Agents Chemother 40*:1754–1756, 1996.
58. A Cingolani, A Antinori, M Sanguinetti, L Gillini, A De Luca, B Posteraro, F Ardito, G Fadda, L Ortona. *J Infect Dis 179*:1025–1029, 1999.

59. V Deretic, E Pagan-Ramos, Y Zhang, S Dhandayuthapani, LE Via. *Nat Biotechnol 14*:1557–1561, 1996.
60. Y Zhang, S Dhandayuthapani, V Deretic. *Proc Natl Acad Sci USA 93*:13212–13216, 1996.
61. J Drews. *Science 287*:1960–1964, 2000.
62. RE Lee, M Protopopova, E Crooks, RA Slayden, M Terrot, CE Barry 3rd. *J Combichem* in Press, 2003.
63. RA Slayden, RE Lee, M Protopopova, CE Barry 3rd. *Antimicrob Agents Chemother,* Submitted 2003.
64. DF Veber, FH Drake, M Gowen. *Curr Opin Chem Biol 1*:151–156, 1997.
65. D Young. *Ann NY Acad Sci 953*:146–150, 2001.
66. JH Grosset. *Rev Infect Dis 11*:S347–S352, 1989.
67. S Sharma, I Verma, GK Khuller. *Eur Respir J 16*:112–117, 2000.
68. D Kaur, GK Khuller. *Int J Antimicrob Agents 17*:51–55, 2001.
69. VA Snewin, MP Gares, PO Gaora, Z Hasan, IN Brown, DB Young. *Infect Immun 67*:4586–4593, 1999.

9

Phenotype MicroArrays: Their Use in Antibiotic Discovery

Barry R. Bochner
Biolog, Inc., Hayward, California, U.S.A.

This chapter describes a new technology, Phenotype MicroArrays (PMs), designed to speed the process of antibiotic discovery and help eliminate drug candidates with unsatisfactory side effects before too much time and money is invested in bringing them to clinical trials. The technology permits scientists, for the first time, to perform broad-based and comprehensive cell testing in an efficient high-throughput format. The technology works uniformly well with a very wide range of microorganisms, including Gram-negative bacteria, Gram-positive bacteria, yeast, and filamentous fungi. Over a dozen important model species have been successfully tested with PMs, including *Escherichia coli*, *Salmonella typhimurium*, *Pseudomonas aeruginosa*, *Staphylococcus aureus*, *Streptococcus pneumoniae*, *Listeria monocytogenes*, *Saccharomyces cerevisiae*, *Candida albicans*, and *Aspergillus nidulans*.

Figure 1 shows a typical drug development sequence. Phenotype MicroArrays can be used in four of the five steps leading up to the decision to put a drug candidate into clinical trials. In addition to their wide applicability, PMs offer unique advantages. The biggest advantages that PMs offer are (1) the ability to provide data at the cellular level, which is often easier to collect and interpret than molecular/biochemical information; and (2) the ability to test drugs with cells under thousands of physiological states. The importance of these advantages is discussed in the sections that follow.

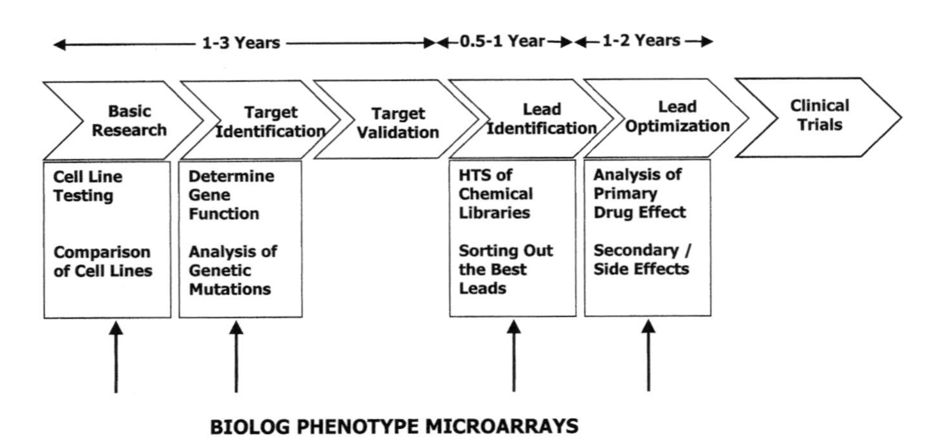

FIGURE 1 The process of drug discovery and development. Phenotype MicroArrays can be used in four of the five stage leading to clinical trials of a new drug, making the process more efficient and productive.

GENERAL DESCRIPTION OF PHENOTYPE MICROARRAYS

Phenotype MicroArrays are arrays composed of thousands of wells, with each well containing a chemistry designed to test a different cellular property—a phenotype (1). The scientist simply adds a cell suspension into the wells, and the array results measure thousands of cellular properties simultaneously.

Phenotype MicroArrays measure a color change as cell respiration results in the reduction of a dye to form a color. The system is flexible in that different detection chemistries and colors can also be used and recorded. Phenotype MicroArrays are designed to scan thousands of metabolic and physiological properties of cells, including chemical structure of the cell envelope, transport functions, energy production, catabolic and biosynthetic pathways, ribosomes, polymerases and other cellular machinery, repair functions, stress responses, and so on.

The first-generation set of PMs tested about 700 phenotypes as follows: 100 carbon sources; 100 nitrogen sources; 100 phosphorus and sulfur sources; 100 tests for biosynthetic pathways; and 300 tests of cell sensitivity to toxic chemicals, including antibiotics, toxic metals, detergents, enzyme inhibitors, etc. (1). The current generation PMs consist of about 2000 phenotypes as follows: 200 carbon sources, 400 nitrogen sources, 100 phosphorus and sulfur sources, 100 tests for biosynthetic pathways, 200 tests for general stress conditions, and 1000 tests of cell sensitivity to toxic chemicals.

USE OF PHENOTYPE MICROARRAYS IN FINDING AND EVALUATING NEW DRUG TARGETS

It has been difficult to find new and unique drug targets and most current antibiotics are variants of about 15 base structures and are aimed at a limited set of targets such as cell wall synthesis, ribosome function, and DNA topoisomerase (2). A common philosophy in hunting for new targets is that they must be products of essential genes and many people are simultaneously pursuing efforts to find and patent this set. Another common assumption is that antibiotics should be broad spectrum in their mode of action.

However, these assumptions may be limiting the scope of thinking and delaying the development of new and better antibiotics. For example, recent work on an interesting new antifungal drug, rapamycin, has shown that it has a new and unique mode of action, complexing with immunophilin proteins (peptidyl-prolyl isomerases) and blocking cell-cycle progression by inhibiting TOR kinases (3). But immunophilins are not essential proteins in yeast (4) so this drug would not have been discovered using screens that only accept essential genes as targets.

Broad-spectrum antibiotics have the advantage that they can be prescribed for a wide range of infections, but along with this comes the disadvantage that they also kill normal microbial flora in and on the body. This can lead to an unacceptable level of side effects and in some cases can worsen the situation by lowering the body's defense barriers. For example, infections involving toxigenic *E. coli* strains such as O157 are often not treated with antibiotics (5). Once the pathogenic cells escape through the lining of the colon, antibiotic treatment may exacerbate the situation by causing release of more toxin and also damaging normal colon flora. Another example is prolonged antibiotic treatment creating conditions that favor the establishment of *Clostridium difficile* in the colon (6).

An alternative approach is to look for antibiotics targeted at "pathogenicity genes" (7). For example, recent genome sequencing of nonpathogenic (8) and pathogenic (9) *E. coli* strains has shown that the pathogens contain 1387 additional genes, many of which presumably code for functions which cause or permit *E. coli* to become an invasive pathogen. A subset of these genes may be excellent new drug targets. To judge this, we need to have a better understanding of the pathogenic process, along with tools to determine gene function.

Phenotype MicroArrays provide a tool for determining gene function. Figure 2 shows the process by which PMs are used to compare two cell lines. Cell suspensions are added to two sets of PMs; the PMs are incubated for 24 to 48 hr; and the OmniLog instrument incubates, monitors, and records the cellular response in all wells of the array. One cell line is recorded as a red tracing and one as a green tracing. To compare these, computer software automatically overlays these tracings and changes areas of overlap to yellow. Thus, in the comparison, phenotypic responses common to both cell lines are yellow, whereas responses unique to one cell line are either red or green.

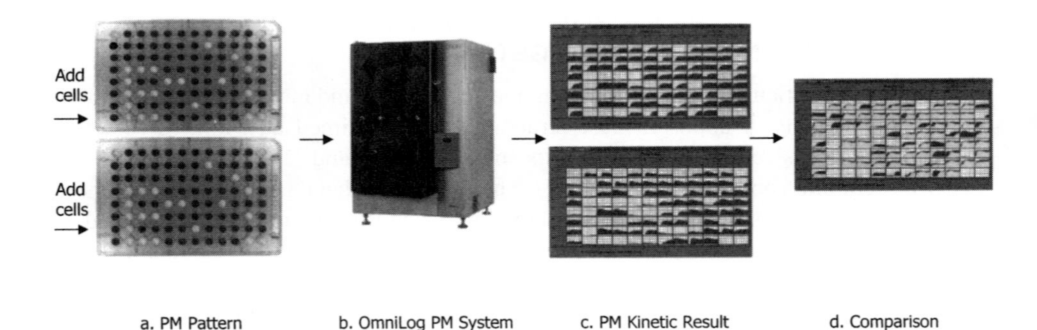

Figure 2 The PM testing process. This example shows two cell lines that differ by one or more genes compared using PM technology. (a) Tests are performed by adding the cells into identical Phenotype MicroArrays. Each cell line produces is characteristic PM pattern, reflecting its phenotypic properties. (b) The PMs are incubated in the OmniLog PM instrument which simultaneously incubates the PMs and monitors the color pattern that develops. (c) PM software records the resulting kinetic phenotypes. One cell line is recorded as a red tracing and the other as a green tracing. (d) The software can generate a variety of comparisons including an overlay of the kinetic phenotypes. Phenotypes common to both strains are colored yellow.

Figure 3 illustrates this with a specific example. In this case PMs measuring the carbon metabolism of the cell are used to compare the nonpathogenic *E. coli* strain MG1655 to the pathogenic strain O157:H7. It has been known for years that O157 is unusual in that it lacks the ability to metabolize sorbitol and it is commonly isolated by clinical labs using Sorbitol MacConkey Agar. The example shown in Figure 3 finds many differences in carbon metabolism between the pathogen and nonpathogen, including the sorbitol defect in O157. Another defect in O157 seen in this example is glucarate metabolism, which has been found missing in about 92% of O157 strains in a comprehensive study covering many strains (10). We also can detect two carbon utilization pathways (for sucrose and adonitol) present only in the pathogen.

Another example to illustrate this point comes from our recent work with the gram-positive pathogen *Listeria monocytogenes*. Using PMs to test isogenic strains kindly provided by Dr. Jose Vazquez-Boland, we have been able to show and confirm that hyperpathogenic strains containing the prfA* allele (11) have picked up the capability to use a variety of sugar phosphates as sources of both carbon and phosphorus (data not shown). This suggests the presence of a sugar phosphate utilization enzyme playing a role in pathogenesis, presumably providing the invading *Listeria* with carbon and phosphorus nutrients for growth while inside the human cell.

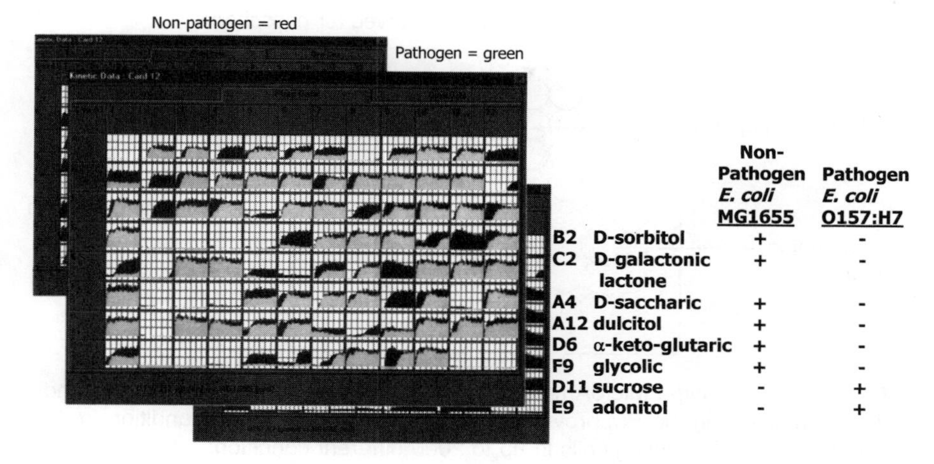

	Non-Pathogen *E. coli* MG1655	Pathogen *E. coli* O157:H7
B2 D-sorbitol	+	-
C2 D-galactonic lactone	+	-
A4 D-saccharic	+	-
A12 dulcitol	+	-
D6 α-keto-glutaric	+	-
F9 glycolic	+	-
D11 sucrose	-	+
E9 adonitol	-	+

FIGURE 3 Comparison of a non-pathogenic and pathogenic bacterial strain. The non-pathogen, *E. coli* MG1655 (red) is compared to the pathogen, *E. coli* O157 (green). Major differences are shown in the table to the right. In carbon metabolism, O157 loses six phenotypes and gains two.

Gain of function is more interesting because in hunting for new drug targets the goal is to find biological processes present only in the pathogen. Of course, to find a truly valid target one would need to examine a large enough set of pathogens and nonpathogens to convincingly show that the phenotype and gene were highly correlated and preferably essential to pathogenicity. Using PM technology this becomes a very efficient and straightforward research project.

Another viable approach is to make gene knockouts of genes found in pathogenicity islands and then attempt to determine their function using PMs. We have recently shown that PMs can be used to determine gene function by direct assay of gene knockouts (1). The process is again as shown in Figure 2, but this time the strains tested are an isogenic pair with the gene of interest knocked out or altered.

PHENOTYPE MICROARRAYS PROVIDE A MORE COMPREHENSIVE MEANS OF TESTING THE EFFECTS OF CHEMICALS ON CELLS

The cell is not a single, static entity. Instead it is more like a multistate automaton capable of a limitless number of states. The physical and chemical environments

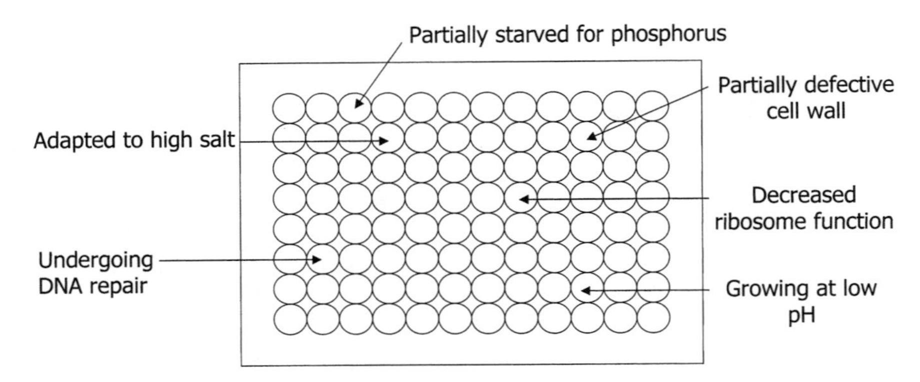

Figure 4 Phenotype MicroArrays grow cells under hundreds of different conditions. Each well in the PM provides a different growth or stress condition for the cell. PMs allow for testing cells in up to 2000 different conditions simultaneously. When the cell grows under a different condition, it changes and adapts, becoming a different cell.

of cells are constantly changing and the cell is constantly monitoring its environment and adapting to it. Every time a cell changes, it changes its gene expression levels, its protein composition, its membranes and receptors, and so on. It becomes a different cell.

A major limitation to current methods for testing the effects of chemicals on cells is that they test the chemical on the cell *under one condition or state of the cell*—usually something approaching optimal growth conditions. This gives us a very narrow and incomplete picture of the potentialities. A simple example is penicillin. Penicillin and other drugs of its class inhibit cell wall synthesis causing cells to lyse when they are in a state of rapid growth. However, these same drugs are ineffective in killing slowly growing or static cells and lysis is also inhibited if the cells are in a hypertonic environment.

Phenotype MicroArrays provide a very simple way to take one cell population and convert it into hundreds or thousands of subpopulations arrayed into diverse physiological states. The intent of the design is to enable a broad survey of the range of possibilities. As diagrammed in Figure 4, in one well the cell may be adapting to low pH, high salt, low phosphorus levels, slowed ribosome function, DNA damage, and so on.

Figure 5 shows how PMs can be used in a simple but powerful testing process to assay the effect of chemicals on cells. Identical cell suspensions are prepared. One suspension is used as a control (no drug) and the other suspensions are dosed with various chemicals to be tested. The level of chemical added is a

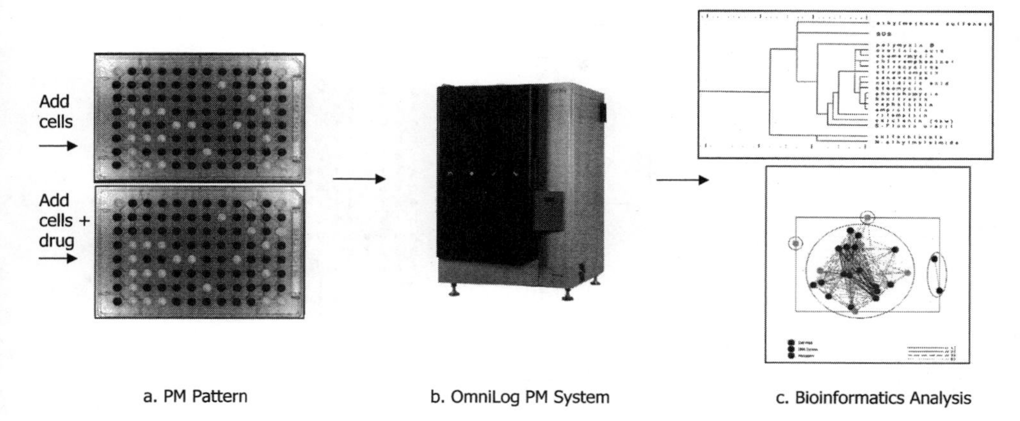

Figure 5 Testing drugs on cells in Phenotype MicroArrays. The testing process for drugs is essentially the same as the process for testing cells with genetic differences shown in Figure 2. In one array, cells alone are added. In a second array, a drug is added to the cells just prior to inoculation into the array. A library of drugs can be tested in this manner, and their PM fingerprints can be compared using clustering algorithms.

slightly inhibitory level (e.g., near the minimum concentration required to give barely detectable growth inhibition). Each suspension is used to inoculate a set of PMs, and again, as in Figure 2, the PMs are incubated, monitored, and recorded by the OmniLog instrument.

USE OF PHENOTYPE MICROARRAYS IN HIGH-THROUGHPUT CHARACTERIZATION OF CHEMICAL LIBRARIES

Depending on the mode of action of the drug, it will be more or less toxic to the cell under certain physiological states of the cell; in other words, in different wells of the PM. Each drug will produce a characteristic "fingerprint" pattern from the PM analysis. Drugs with similar modes of action will have similar "fingerprints" and vice versa. The fingerprints can then be sorted or grouped using standard clustering and/or pattern recognition algorithms.

Figure 6 shows an example with data collected on a set of 20 chemicals tested against *E. coli*. The dendrogram clustering method accurately grouped four drugs that are cell wall inhibitors, three drugs that are ribosome inhibitors, and two of three drugs that are DNA topoisomerase inhibitors. Polymyxin B, ceru-

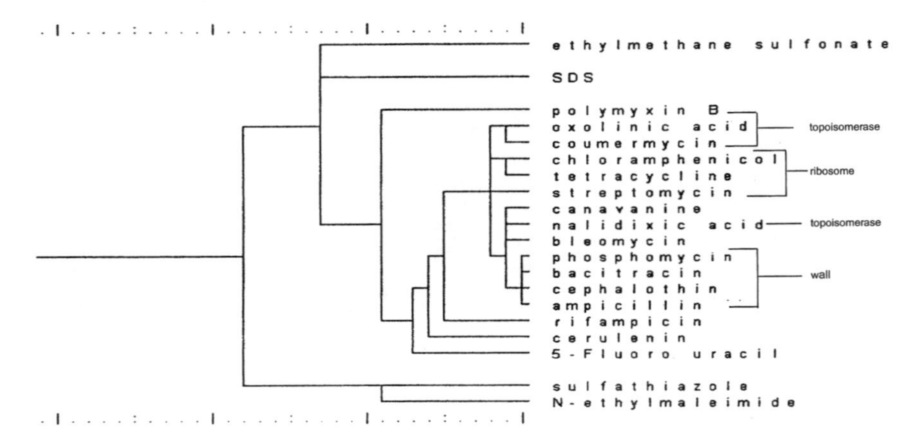

FIGURE 6 Fingerprinting the drug sulfamethoxazole using PMs. *E. coli* bacteria were tested without or with sulfamethoxazole added. A decrease in well color indicates relative sensitivity (synergy) with a second drug in the PM well. An increase in well color indicates relative resistance (antagonism).

lenin, and rifampicin, which have unique modes of action, branch off as unique patterns. Furthermore, this analysis clearly differentiated the antibiotics having a more targeted mode of action from the chemicals that are generally toxic (e.g., reactive alkylating agents such as ethylmethane sulfonate and *N*-ethyl maleimide and denaturants such as sodium dodecyl sulfate).

A major activity in antibiotic development is screening chemical libraries. Phenotype MicroArrays show the prospect of providing a highly efficient and comprehensive testing technology. Thousands of chemicals can be "fingerprinted" in this approach. First, chemicals with general toxicity can be eliminated from consideration. Second, chemicals can be grouped by mode of action. This allows the sorting of chemicals into groups such as (a) identical to known antibiotics, (b) similar to but distinct from known antibiotics, or (c) clearly distinct from known antibiotics and potentially unique new leads.

Phenotype MicroArray fingerprints are highly reproducible, so databases can be compiled and compared with data from chemical libraries screened at a later time. The end result is an expandable database that should give stronger conclusions and better insights as it grows in size. To characterize a library, chemicals would only need to be fingerprinted once unless there was an important reason to test them again with a different model cell system. This is much more efficient than current methods whereby chemical libraries are rescreened in cell-based assays each time a new target is evaluated.

USE OF PHENOTYPE MICROARRAYS TO DETECT SYNERGIES AND ANTAGONISMS WITH KNOWN DRUGS

As briefly mentioned above, many of the PM wells contain known drugs. Drugs are useful as probes in the PM wells because they selectively attack certain cellular targets. When, as diagrammed in Figure 5, drugs are added to PMs that already contain other drugs, we are provided with data on thousands of drug combinations. Therefore this has the potential to provide data on drug synergy and antagonism.

An example is shown in Figure 7. In this case we are testing the effect of sulfamethoxazole on *E. coli* using a PM with 24 antibiotics. Sulfamethoxazole is known to be synergistic with trimethoprim (12) and this is confirmed by the PM result in which there is a decrease in color in the trimethoprim well when sulfamethoxazole is present. There is also an apparent synergy of cephaloridine with sulfamethoxazole. On the other hand, the same array shows antagonism with novobiocin, tetracycline, and chlortetracycline.

With our limited knowledge of cell physiology, it is usually not obvious why certain drugs are synergistic or antagonistic. However, it can be very valuable to detect these interactions by the simple PM assay process. A drug lead might be overlooked because it has little effect by itself, yet it might have a strong synergy with another drug making the combination valuable enough to consider clinical trials. Conversely, a drug lead may look promising on its own, yet in PM assays it proves to have adverse interactions with another drug that could be taken at the same time. Adverse drug interactions would normally be determined

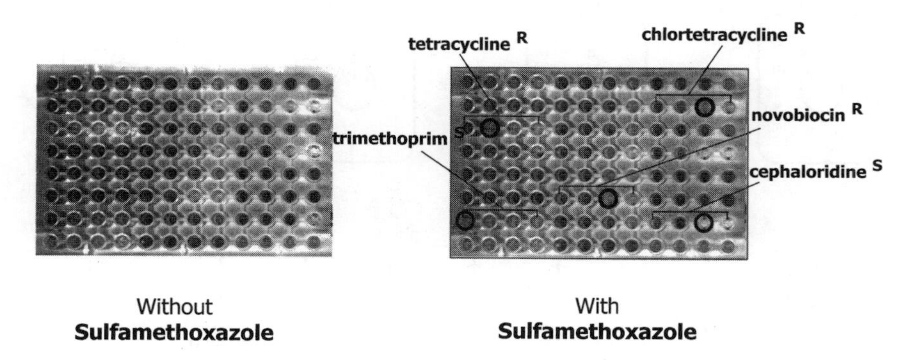

FIGURE 7 Effects and side effects of drugs shown by PM testing. Drug 1 is targeted against protein 1 in the cell and it gives the same phenotypic PM pattern (loss of phenotype 1) as the mutant cell line with protein 1 inactivated genetically. Drug 2 is also targeted against protein 1, but it produces a side effect, changing phenotype 5 as well as phenotype 1.

at a later step in the drug development process as part of additional toxicological testing. With PM testing, drug interaction is determined early on and without extra cost, as an integral part of the process.

USE OF PHENOTYPE MICROARRAYS TO DETECT SECONDARY AND SIDE EFFECTS OF DRUG CANDIDATES

Cell-based assays are commonly utilized to test the effect of potential drug leads on a cellular target. However, these assays provide no information about how these leads might also affect other unintended cellular targets. This is key information because it is the action on "other" cellular targets that produces side effects. Phenotype MicroArrays provide the first straightforward approach to address this major issue.

Figure 8 shows how PM assays should detect side effects in a comprehensive way. In the cartoon shown, both drug 1 and drug 2 hit the target protein, protein 1. Since protein 1 is involved in phenotype 1, the effect of the drugs in inhibiting protein 1 is manifested at the cellular level as a change in phenotype 1. However, drug 2 has a second effect; it also inhibits protein 5. In the PM assay of drug 2 this is manifested and detected as a change in phenotype 5 as well as phenotype 1. A scientist needing to choose between drug 1 and drug 2 would therefore have a sound basis for choosing. If the drugs were targeted to human

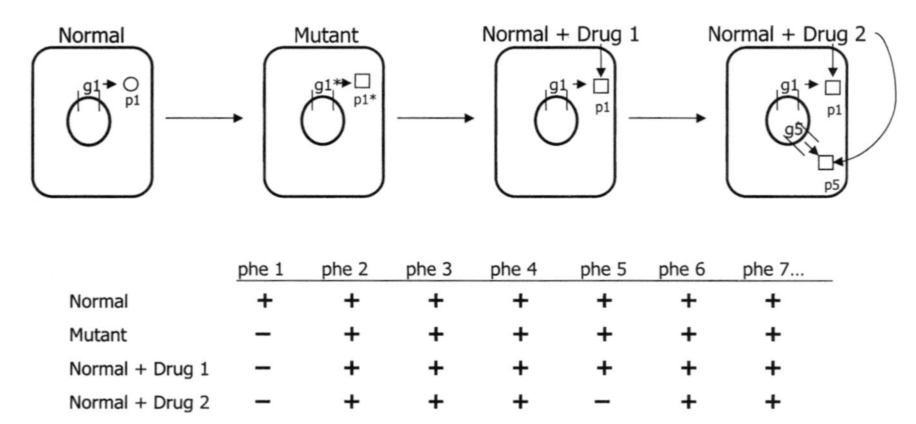

	phe 1	phe 2	phe 3	phe 4	phe 5	phe 6	phe 7...
Normal	+	+	+	+	+	+	+
Mutant	–	+	+	+	+	+	+
Normal + Drug 1	–	+	+	+	+	+	+
Normal + Drug 2	–	+	+	+	–	+	+

FIGURE 8 "Binning" of chemicals toxic to *E. coli* based on PM patterns. PM patterns for twenty chemicals were recorded and analyzed using a clustering algorithm. Nearly all of the chemicals with similar modes of action were clustered within the same dendrogram groups.

cells and side effects were considered undesirable, drug 1 would be a better choice to pursue. Conversely, if the drugs are antimicrobials, drug 2 may be preferred since it has two sites of action, making it more difficult for cells to become resistant.

Here again, it is a major advantage that PMs can test the effect of a chemical under thousands of states of the cell. Standard methods for cell-based assay would only look for potential side effects or secondary targets on cells undergoing rapid growth. The PM assay format extends the view into thousands of physiological states of the cell. A drug side effect may only come into play in cells stressed in a certain way. Our body is made up of many cell types in a range of physiological states.

Phenotype MicroArrays provide an attractive and very practical approach to the difficult challenge of detecting secondary and side effects early in the drug development process.

SUMMARY AND CONCLUSIONS

Cell-based testing is already widely used as an integral part of antibiotic drug discovery. Typically, cell-based assays are used in screening chemical libraries to find the subset of chemicals that "hit" the target, to determine drug mode of action, to measure and characterize the rate and nature of drug resistance, and in many toxicological assays. However, cell-based assays have not been used as a general tool for "fingerprinting" chemicals or for systematically testing for potential side effects.

Phenotype MicroArrays are a promising new tool with many applications in antibiotic drug discovery. They can be used to determine entirely new and novel targets for drugs or to evaluate genes that are already thought to hold promise. Drug candidates can also be evaluated in PMs. They can be tested, in high-throughput mode, to sort them by mode of action, to determine synergy/antagonism with other drugs, and to determine potential secondary and side effects. All of these results are obtained simultaneously in a single analysis.

Phenotype MicroArray technology provides essential information to eliminate unsatisfactory drug candidates before costly clinical trials are initiated. Pharmaceutical and biotech companies do not waste money on drugs that will fail but instead are able to focus resources on those that will be successful. This should also allow them to dramatically improve the efficiency of the drug discovery process to bring better drugs to the market at a faster pace.

REFERENCES

1. BR Bochner, P Gadzinski, E Panomitros. *Genome Res 11*:1246–1255, 2001.
2. H Breithaupt. *Nat Biotechnol 17*:1165–1169, 1999.

3. MC Cruz, LM Cavallo, JM Gorlach, G Cox, JR Perfect, ME Cardenas, J Heitman. *Biology 19*:4101–4112, 1999.
4. K Dolinski, S Muir, M Cardenas, J Heitman. *Proc Natl Acad Sci* USA *94*: 13093–13098, 1997.
5. PS Mead, PM Griffin. *Escherichia coli* O157:H7. *Lancet 352*:1207–1212, 1998.
6. SD Allen, CL Emery, JA Siders. *Clostridium*. In: PR Murray, EJ Baron, MA Pfaller, FC Tenover, and RH Yolken, eds. *Manual of Clinical Microbiology*, 7th edition. Amer Society for Microbiology, 1999.
7. S De Baets, K Mortelmans. *SIM News 51*:125–131, 2001.
8. FR Blattner et al. The complete genome sequence of *Escherichia coli* K-12. *Science 277*:1453–1474, 1997.
9. NT Perna et al. Genome sequence of enterohaemorrhagic *Escherichia coli* O157: H7. *Nature 409*:529–533, 2001.
10. GR Siragusa, *ASM Abstr P-49*:521, 1999.
11. JA Vazquez-Boland, M Kuhn, P Berche, T. Chakraborty, G Dominguez-Bernal, W Goebel, B Gonzalez-Zorn, J Wehland, J Kreft. *Clin Micro Rev 14*:584–640, 2001.
12. GH Hitchings, JJ Burchall. *Adv Enzymol 27*:417–468, 1965.

10

Microbial Proteomics: New Approaches for Therapeutic Vaccines and Drug Discovery

C. Patrick McAtee
Lexicon Genetics, Inc., The Woodlands, Texas, U.S.A.

> . . . There is a man working here—his name is Pasteur—who is finding out wonderful things about the machinery of life . . . he is even going to find out, perhaps, what causes disease! . . .
>
> —Student in the Ecole normale superiere, Paris, 1861

Contrary to what some pharmaceutical "visionaries" have predicted, pathogenic microbes have not gone away, but persist as a significant problem in modern medicine. Thanks to years of inappropriate antibiotic therapies, they're back with a vengeance. Antibiotic-resistant pathogens have emerged and continue to wreak havoc on public health worldwide. The common denominator in this developing scenario is the ever-evolving genomes of microorganisms. There are many approaches suitable to monitoring fluctuations in gene expression and microbial adaptation, which have been presented in numerous review articles. However, of these various genomic approaches, the proteomic approach is unique in that a single genome can give rise to qualitatively and quantitatively different proteomes. The term *proteome*, overused as it is, represents the expressed protein complement of the genome. By examining pathogens at the protein level, one can monitor phenomena such as protein expression levels, posttranslational modifications and processing, as well as study multisubunit protein complexes. As I work with proteins for a living, I would, like to believe that proteomics can do it all. However, the reality is that proteomics is really nothing new but simply a newly coined moniker that represents analytical biochemistry with newer, neater equipment. Many of the technologies that make up what is now the current buzzword, "proteomics," have been in existence for some time. The key to obtaining useful information from the "new" field of proteomics lies in an older, stodgier field—that of cell biology.

TECHNOLOGY—FROM GELS TO MASS SPECTROMETERS

While the characterization of protein structure and function has long been a traditional part of biochemistry, the advent of modern mass spectrometry techniques, in conjunction with advances in DNA sequence databases and bioinformatic analysis, has certainly given a boost to the field of proteomics. However, every analytical approach has both its advantages and pitfalls. Granted, through proteomics, previous impediments to protein target identification such as degree of purity and the necessity for relatively high quantities of expressed protein have in essence been eliminated. Using older technologies such as 2D electrophoresis, enhanced technologies such as new-generation mass spectrometers, and novel technologies such as genomic sequence databases, proteomics or what some would call "real protein biochemistry" has come a long way in a brief period of time.

The beginnings of microbial proteomics can be traced back to the introduction of two-dimensional electrophoresis by O'Farrell, in which proteins were separated initially by isoelectric focusing in low-percentage acrylamide poured in tubes followed by molecular weight separation using sodium dodecyl sulfate – polyacrylamide gel electrophoresi (SDS–PAGE) in the second dimension (1). Nearly every graduate student, medical fellow, and post-doc of this era should have fond memories of the many times that an overnight focusing was initiated only to find that the gel had separated from the tube in the wee hours of the morning. Thus the buffers poured into each other and the power source simply gave up. For those of us studying phosphorylation, these frustrations were usually compounded by the creation of a sizable pool of phosphorus-32 on our benchtops. Further, it was often difficult to remove tube gels from the tube if the run was actually successful. Invariably, unless one had the dexterity of a senior neurosurgeon, the flimsy tube gels would get chopped up into a million pieces or they would fly out of the tubes and never be found (unless it was a kinase study and then a good-quality Geiger counter usually picked up the trail). However, when tube gels worked properly, they were very good and the power of 2D electrophoresis as a biochemical separation technique was firmly established as O'Farrell was able to resolve 1100 different proteins from *Escherichia coli.*

Improvements to the O'Farrell approach came later in the form of immobilized pH gradients, which significantly enhanced the reproducibility of 2D electrophoresis (2). Comparison of the spotting patterns on 2D electrophoretic profiles can allow for the quantification and identification of modified and up- or downregulated proteins in a biological sample. Then, of course, there was the identification of these 1100 "spots", which at the time would have been a pretty heroic accomplishment (and still would be no small feat!). Now, however, the spot may be extracted, digested with an endoprotease such as tryspin or Lys-C, and subjected to mass spectrometry in a rather direct and unencumbered manner. The experimental peptide masses may then be compared to the theoretical masses (calculated from the genomic information) through a variety of programs such as MASCOT and MS-Fit and a match is obtained (3,4).

Yates and coworkers, utilizing a program called SEQUEST, developed further refinement of this database-searching procedure (5). SEQUEST correlates tandem mass spectra of peptides with individual amino acid sequences from protein and nucleotide databases (see Figure 1). While the instrumentation and analytical programs have become significantly refined with reported sensitivities down to the femtomole level, most analysts agree that the 2D gels appear to be the rate-limiting factor in analysis. Another area of concern is that the dynamic ranges of electrophoresis-analytical software have often lead to publications reporting extraordinary numbers of spots isolated from a gel that in reality were

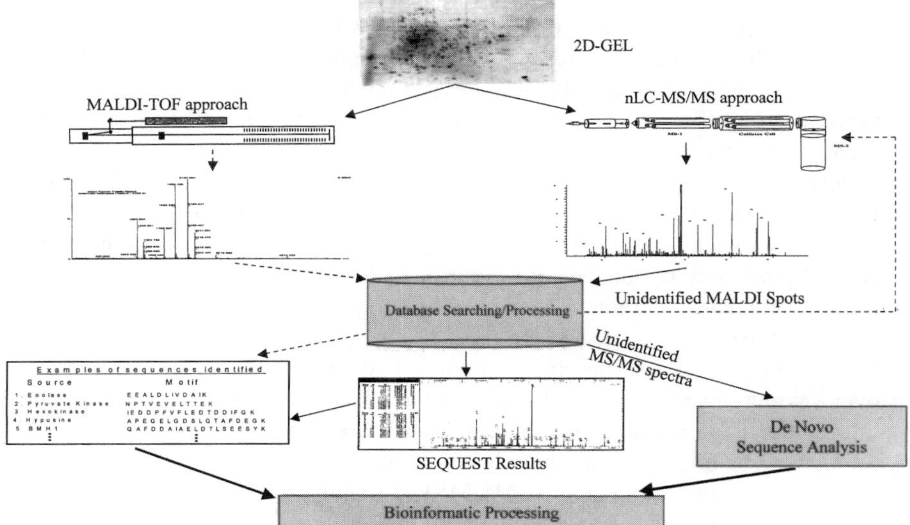

FIGURE 1 The Wonder of Technology: Proteomics in a nutshell. A sample is run on a two-dimensional gel and the resulting protein bands may be extracted, digested with a specific endoprotease, and then subjected to mass spectrometry. Using MALDI-TOF or nanospray LC-MS/MS, spots may be identified using database searching. Spots not identifiable through MALDI-TOF and typical mass analysis programs may then be subjected to nLC-MS/MS and identified through SEQUEST analysis.

either instrumentation limitations or other artifacts. Reproducibility can also be an issue when running replicates of the same sample.

An alternative to the much improved, but still laborious, 2D gel approach is that of 2D liquid chromatography – mass spectrometry (LC-MS) in which a protease digested sample is simply injected directly onto a column and has been described by Yates and termed multidimensional protein identification technology (MudPIT) (6). In this procedure, proteins are digested in solution and the peptide mixture is loaded onto a strong cation-exchange column; peptides are sequentially eluted with increasing electrolyte concentration according to their charge and hydrophobicity. Each ion-exchange fraction is then subsequently separated by conventional reverse-phase liquid chromatography combined into a biphasic column with reverse-phase and strong cation-exchange material packed sequentially in the same column. The eluent is subjected to direct mass spectrometric analysis. This approach is limited by a slow sample-loading rate. Also, it is not possible to quantitate information much less get a "global view" of what is going on in the proteome by this approach.

An attempt to resolve the issues of quantitation in LC-MS has been described by Aebersold and colleagues (7). Described as a method that has the potential to identify and quantify most, if not all, the proteins expressed by a cell or tissue, it is based on a new class of reagents termed isotope-coded affinity tags (ICAT) (8). The ICAT reagent consists of a biotin affinity tag, a polyether linker that can incorporate stable isotopes (i.e., deuterium), and an iodoacetamide-reactive group that specifically reacts with cysteinyl thiols. Stable isotopes are incorporated into proteins by selective alkylation of cysteines with either an isotopically heavy or light reagent. The proteins in one sample are labeled with the isotopically heavy light reagent and the proteins in the second sample are labeled with the isotopically heavy reagent. The two protein mixtures are then combined. At this point, any optional fractionation technique can be performed to enrich for low-abundance proteins or to reduce the complexity of the mixture, while the stable isotope tag imprinted in the proteins strictly maintains the relative quantities. Prior to analysis, the protein mixture is digested with trypsin and passed over a avidin–agarose column. Because the ICAT label contains the stable isotope information as well as a biotin tag, ICAT-labeled peptides are selectively isolated for analysis by microcapillary liquid chromatography electrospray ionization tandem mass spectrometry (LC-ESI-MS/MS). If required by the complexity of the sample, the peptide mixture can optionally be further fractionated prior to mass spectrometric analysis. The ratio of ion intensities from coeluting ICAT-labeled peptide pairs of identical sequence precedes accurate quantitation, while a subsequent tandem MS analysis identifies the proteins in the sample by their amino acid sequence. The method is reported to be capable of detecting proteins of very low abundance and is an automatable process. This technology has its own limitations (such as being cysteine specific). In addition, high-quality quantitation

(CVs less than 20%) from LC/MS/MS data, where the peptide pairs do not entirely coelute, has its challenges as well as limited sensitivity. Moral of the story: Damn gels—can't live with them, can't live without them.

YOU WANT CHIPS WITH THAT?

Last, but not least, in the technology toolbox is the so-called proteomics on a chip. Obviously, a critical need exists for the development of novel technologies that will enable high-throughput studies of the proteins encoded by the approximately 30,000–100,000 genes encoded by the human genome. "Traditional proteomics" methodologies, while adequate for basic laboratory approaches, are inadequate for the detection and characterization of large sets of protein–protein interactions. Protein microarray technology recently developed by several labs utilizes protein microchannel microarrays to characterize antibody profiles in a variety of infectious and other disease states (9). Using microfabrication technologies borrowed from the microelectronics industry, these systems offer the ability to realize on-chip microchannel networks that can be computer controlled to transport, mix, separate, and detect protein fragments (10). The advantage is that an automated series of reactions can be performed on 1 to 1,000-pl samples with all reactants and products of the assay remaining suspended in a buffered solution at all times. This is in contrast to the solid-phase detection systems of microarrays typically used in combinatorial proteomics research, where the small spot sizes dictated by the needs for high sample density result in rapid evaporation of solvents. Other purported advantages of the microchannel approach include the ability to combine assays with simple binding event detections as well as the ability purify candidate species of known reactivity for subsequent testing off-chip.

Enough, however, with the "wonders of technology." What is the application of proteomics to disease and, in particular, infectious disease? There are three areas of antimicrobial research that come to mind where proteomics may be able to provide a technological advantage to the discovery of new therapeutics: vaccines, thereapeutics for resistant organisms, and cell-based screening. In the following three "real-world" applications, the opportunities for proteomic approaches are discussed.

REAL-WORLD APPLICATION 1: DETECTION OF VACCINE ANTIGENS IN *HELICOBACTER PYLORI*

Helicobacter pylori is a gram-negative bacterium that chronically infects the gastric mucosa of more than half of all humans worldwide and is a major cause of gastritis and peptic ulcer disease and an early risk factor for gastric cancer (11). Only some 10 to 20% of infections, however, result in overt disease. DNA typing has established that *H. pylori* is extremely diverse as a species, and it is likely that the varied outcomes of infection reflect differences in bacterial genotype, human host genotype, and physiologic, immunologic, and environmental factors (12). These considerations make it valuable to thoroughly characterize the proteins and other antigens that *H. pylori* produces and the human responses to them.

There is a great need for an effective anti-*H. pylori* vaccine, especially in Third World, high-risk populations where *H. pylori* eradication by standard antimicrobial therapies is often followed by reinfection. Much attention has been focused on urease-based vaccines because of the essentiality of urease and some encouraging results with mouse *Helicobacter felis* models. VacA has also been considered a candidate based on results with a mouse *H. pylori* model (13,14). These mouse models, however, may not adequately mimic the human condition and clinical trials of urease vaccines have been only marginally encouraging. This reinforces the sense that other or additional antigens may be needed for a truly effective vaccine.

Other factors important for *H. pylori* colonization or virulence have been identified. Some of the more prominent factors include(1) flagellae (15), which allow the organism to move in the mucous layer;(2) the previously mentioned urease complex (16), which may help maintain a neutral micro pH environment in the face of gastric acidity; (3) the VacA protein (17), which generates vacuoles in eukaryotic epithelial cells; (4) an adhesin binding to the Lewis blood antigen group (18); and (5) the *cag* pathogenicity island (19,20), some of whose encoded proteins help trigger severe inflammatory responses and which, like VacA toxigenicity, is disease associated. Based on the presence of the *cag* pathogenicity island (PAI), the *H. pylori* isolates are subdivided into two types. Type I strains, containing the *cag* PAI, exhibit increased virulence, since they are predominantly associated with severe gastric disease, whereas type II strains, lacking the *cag* PAI, are more frequently isolated from asymptomatic carriers (21). It has been demonstrated that some of the proteins encoded by the *cag* PAI trigger severe inflammatory responses in the host (22). However, the precise function of the gene products of the *cag* PAI and their role in virulence remain to be elucidated.

Several other *H. pylori* proteins with known activities, or which are related to similar proteins of known function in other organisms, have been isolated (23,24). While the complete genomic DNA sequence of *H. pylori* 26695 has been reported, many of the proteins inferred from this DNA sequence have no known function, and this DNA sequence clone does not always predict which open reading frames are likely to encode virulence factors or antigens suitable for diagnostic or vaccine studies.

A number of studies have begun to address associations of specific *H. pylori* antigens to antibodies in patients with particular gastroduodenal pathologies and of possible autoimmune components to *H. pylori*-associated disease. There is very little information, however, regarding the long-term evolution and clinical implications of these human responses before and after the eradication of *H. pylori* by antibiotic treatment regimens. Pharmaceutical therapy to treat the *H. pylori* infection involves expensive combinations of various antibiotics, proton pump inhibitors, and bismuth compounds but shows only a limited efficacy (of approximately 80 to 90%) and does not prevent reinfection after successful eradication. In addition, *H. pylori* strains resistant to the most potent antibiotics used in the treatment of *H. pylori* infections, metronidazole and clarithromycin, are emerging rapidly (25). Considering further that the number of infected people worldwide requiring treatment is far beyond the reach of the antibiotic triple therapy, development of a vaccine seems to make sense for the global control of *H. pylori* infection. It has been shown by various researchers that in animal models of infection protective immunity can be achieved by the coadministration of an appropriate mucosal adjuvant and various *H. pylori* antigens, either separately or in combination, via the orogastric route. The protective antigens identified include the urease; VacA; CagA, the immunodominant marker protein for the presence of the *cag* PAI; catalase; and HspA and HspB, the *H. pylori* homologs of the heat shock proteins GroES and GroEL (26–29). In particular, the *H. pylori* urease gave rise to a high degree of protective immunity in vaccinated animals, and it was reported that 100% protection in *H. pylori*-challenged mice could be achieved by the administration of urease via a live carrier *Salmonella* strain expressing recombinant *H. pylori* subunits A and B (30). Furthermore, as mentioned previously, it has been demonstrated that therapeutic vaccination with recombinant VacA and CagA eradicates a chronic *H. pylori* infection in mice, demonstrating that the inability of the natural immune response to clear *H. pylori* infection can be overcome (31).

Considering the advantage of an efficacious vaccine, it is important to identify the *H. pylori* proteins that elicit a strong immune response in humans in order to analyze their capability to confer protective immunity. Furthermore, the identification and characterization of immunodominant proteins will contribute to the improvement of serological tests for detecting and monitoring *H. pylori* infections. Another important question is whether there exists a correlation between the presence of antibodies directed against specific *H. pylori* antigens and the particular *H. pylori*-associated gastroduodenal pathology from which a patient is suffering. In a previous study, proteomics was used to identify common patterns of *H. pylori* antigens that are recognized by sera from patients showing various gastroduodenal pathologies (32). The proteins from lysed cell pellets of *H. pylori* ATCC 43504 were separated on a series of 2D gels run in parallel with an initial pH gradient of pH 4 to pH 8 (see Figure 2). The silver-stained gel (Figure 2A) revealed prominent individual proteins, with several protein "families"-most notably as clusters of bands at approximately 89 (p*I* 6.8), 66, and 58 kDa (p*I* 6.5). The proteins from these 2D gels were transferred to PVDF membranes and incubated with a positive serum pool (Figure 2B) or a negative serum pool (Figure

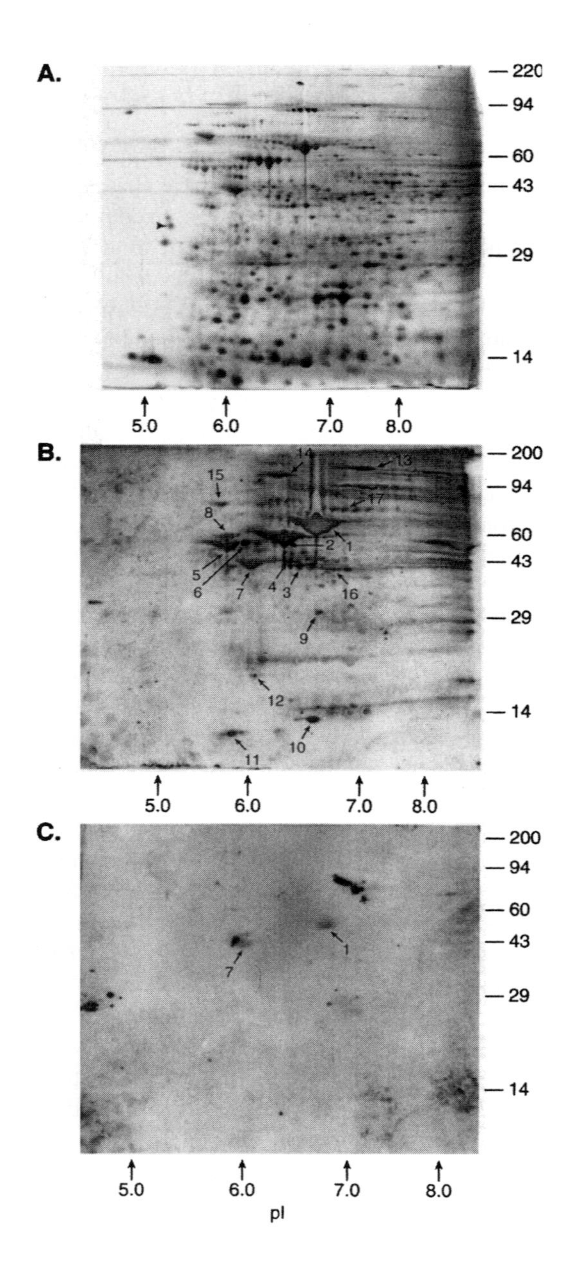
A.
— 220
— 94
— 60
— 43
— 29
— 14
5.0
6.0
7.0
8.0
B.
— 200
— 94
— 60
— 43
— 29
— 14
14
13
15
17
8
2
1
5
6
7
4
3
16
9
12
10
11
5.0
6.0
7.0
8.0
C.
— 200
— 94
— 60
— 43
— 29
— 14
1
7
5.0
6.0
7.0
8.0
pI

2C). Western blot data revealed at least 17 spots or groups of spots that were recognized by antibodies in an infected patient serum pool. Transblotted 2D spots from the pH 4 to 8 gel were sequenced by Edman-type amino acid analysis, with the protein within selected spots evaluated further for internal sequence information. The sequences from these spots were compared with sequences in available databases. Briefly, spots 1 and 2 corresponded to the *H. pylori* urease b subunit and the urease b-associated chaperonin GroEL (33,34), respectively. Spot 3 consisted of two proteins: the major species was pyruvate flavidoxin oxidoreductase (35), and the minor protein species corresponded to the previously described *H. pylori* hypothetical protein 2, or HP0154 (36). Spot 4 corresponded to HP0537, from the *cag* region (37). Spots 5, 6, and 8 (38–40) corresponded to flagellin proteins. Spot 7 consisted of two proteins that did not match any previously reported sequences from *H. pylori*. The major component, however, has 90% homology with the *Escherichia coli* TufB protein (possibly HP1205) (41), and the minor component has some sequence homology with various ATPase proton pumps (42). Spot 9 was homologous to monomine oxidase from various species (43). Spot 10 corresponded to the neutrophil-activating protein (44). Spot 11 corresponded to HP1199, a ribosomal protein (45). Spot 12 had homology with the ClpP protease from various bacteria (46). The sequencing signals of spots 13 and 14 were too low to be read with confidence. Spots 15 and 16 (major) corresponded to HP0109 (Hsp 70) and HP0589 (ferrodoxin oxidoreductase), respectively (47,48). Spots 16 (minor) and 17 corresponded to a protein previously isolated by O'Toole et al. (49). In the control blot with sera from *H. pylori*-negative persons, only the urease b subunit (spot 1), likely due to cross-reaction with ureases of intestinal bacteria, and the spot 7 proteins showed cross-reactivity. Additional unique proteins were found in a nonequilibrium focusing gel followed by SDS–PAGE, even though fewer proteins, overall, were resolved (see Figure 3). Spots 1 through 4 were present in very low quantities; therefore, a clear N-terminal sequence could not be determined with confidence. Spot 5 was the urease b subunit also seen in the pH 4 to 8 2D gels. Likewise, spots 6, 7, and 8 corre-

◀———

FIGURE 2 *H. pylori* 2D map (pH gradient electrophoresis, pH 4 to 8). 200 μg of protein extract was loaded in the first dimension. Molecular size markers are indicated on the right (in kilodaltons). (A) Silver-stained 2D gel. Fifty nanograms of tropomyosin was added as an internal IEF standard. This protein migrates as a doublet with a polypeptide spot of 33 kDa and pI 5.2. (B) Western blot of a duplicate 2D gel with an *H. pylori*-positive serum pool. (C) Western blot of a duplicate 2D gel with an *H. pylori*-negative (control) serum pool. (Figure reprinted with permission from ASM)

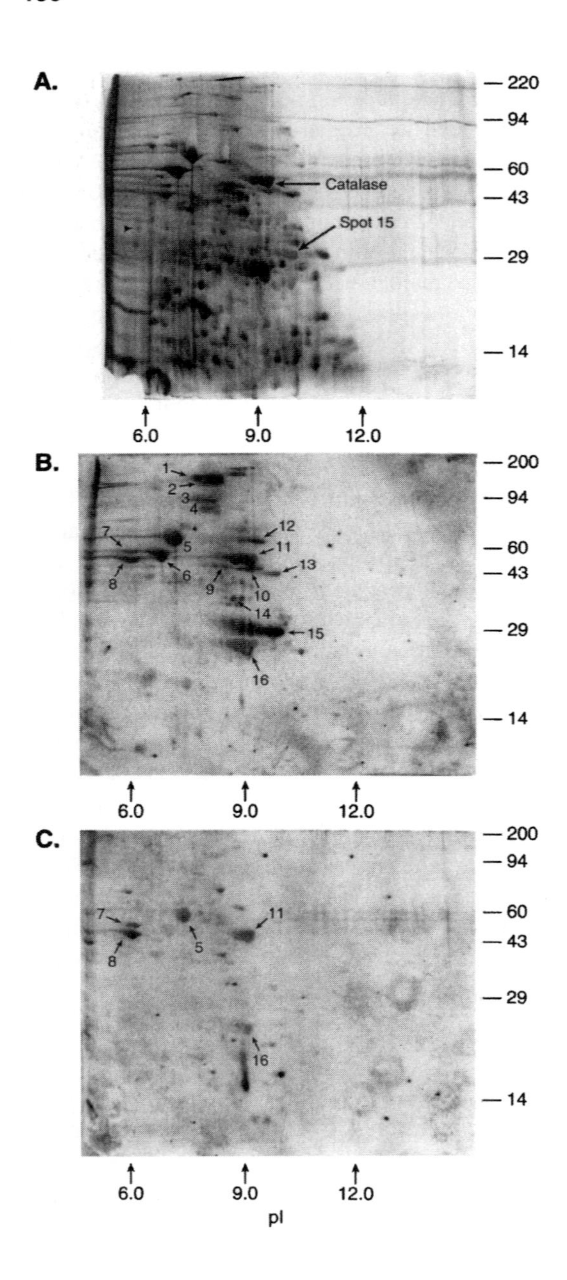
A.
220
94
60
43
29
14
Catalase
Spot 15
6.0
9.0
12.0
B.
200
94
60
43
29
14
1
2
3
4
5
6
7
8
9
10
11
12
13
14
15
16
6.0
9.0
12.0
C.
200
94
60
43
29
14
7
8
5
11
16
6.0
9.0
12.0
pI

sponded to urease b-associated chaperonin, flagellin b precursor, and flagellin a protein, respectively, which were also separated on the pH 4 to 8 2D gel. Spot 9 (major) corresponded to HP0027 (isocitrate dehydrogenase) (50), with spot 9 (minor) representing a possible contaminant in the sequencing sample. Spot 10 corresponded to an open reading frame from HP1018, an open reading frame with no known database homologs (51). Spot 11 corresponded to *H. pylori* catalase (52). Spot 12 contained an N-terminal sequence which has been found in several Omp's (Omp 5, 8, 9, 19, and 27) (53). Spot 13 corresponded to HP1350, a putative protease Spot 14 was the previously reported HopC protein (54), and spot 16 was the urease a subunit (55). The sequence yields from transblotted spot 15 were low (in the midfemtomole range), suggesting that the protein was blocked. The sequence information derived from spot 15 gave an N-terminal amino acid sequence that did not match any known protein sequences. This suggested that the protein(s) in this spot might be modified at the amino terminus, as sequencing yields were low despite the protein(s) being clearly visible on a silver-stained gel. These experiments not only illustrate that 2D gel electrophoresis can give a global view of the abundant proteins of *H. pylori*$_1$ but also suggest that the identification of large numbers of proteins and their characterization with defined serum pools raises the possibility of rapid screening for potential vaccine (as well as diagnostic) candidates. Amino-terminal sequencing and/or proteolytic mass spectral mapping on isolated spots allowed for efficient characterization of these potential antigens. Using this approach upfront, downstream peptidomimetic analysis in parallel with libraries of cloned DNA fragments could provide additional information for the construction of specific vaccine clones or diagnostic recombinant "mosaic" antigens (see Figure 4). This is especially important in the case of pathogens whose genomes have not yet been sequenced. One of the many advantages in using "proteome"-type technologies, as here, as opposed to traditional molecular biology (DNA) library approaches, stems from information about likely functionality and utility that comes from initial screening and that is refined as candidate antigens are discovered.

Figure 3 *H. pylori* 2D map (nonequilibrium pH gradient electrophoresis, pH 8 to 13). 200 μg of protein extract was loaded in the first dimension. Molecular size markers are indicated on the right (in kilodaltons). (A) Silver-stained 2D gel. Fifty nanograms of tropomyosin was added as an internal IEF standard. This protein migrates as a doublet with a polypeptide spot of 33 kDa and pI 5.2. Purified lysozyme (14 kDa, pI 10.5 to 11.0) was also added as an internal pI standard. (B) Western blot of a duplicate 2D gel with an *H. pylori*-positive serum pool. (C) Western blot of a duplicate 2D gel with an *H. pylori*-negative (control) serum pool. (Figure reprinted with permission from ASM)

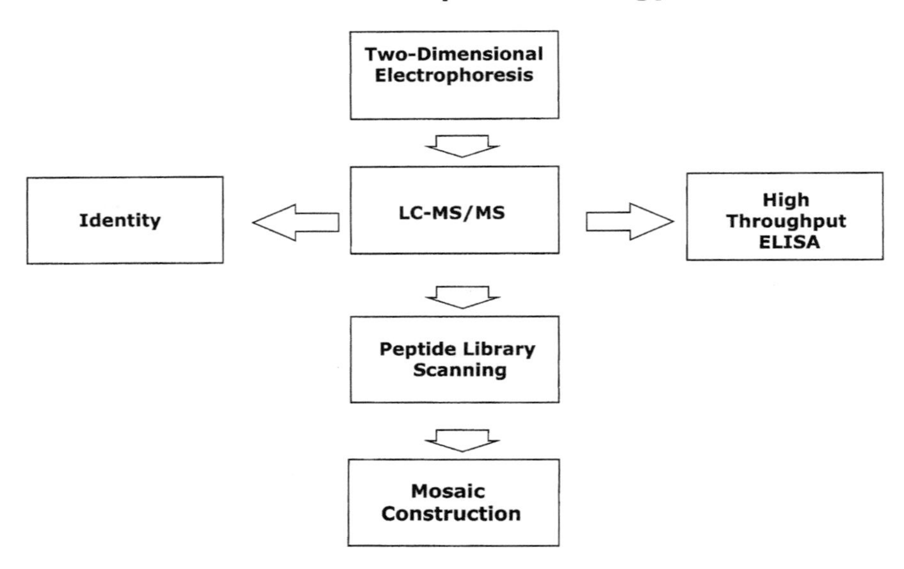

FIGURE 4 Vaccine development strategy. Proteomics in combination with library screening approaches can enable for the simultaneous identification and characterization of potential vaccine candidates. Following screening with serum panels, epitope scanning can be carried out and precise antigenic determinants mapped. A recombinant mosaic vaccine can then be constructed based upon these multiple epitopes.

REAL-WORLD APPLICATION 2: IDENTIFICATION OF DIFFERENTIALLY REGULATED PROTEINS IN METRONIDAZOLE-RESISTANT *HELICOBACTER PYLORI*

As described, *H. pylori* is a microaerophilic bacterium that exhibits a strict respiratory form of metabolism, which therefore restricts it to an environment of relatively low oxygen tension, such as the gastric mucosa. Microaerophiles are susceptible to redox-active 5-nitroimidazole drugs that are converted to short-lived, DNA-damaging hydroxylamine derivatives by low-potential reductases, including ferrodoxins and flavidoxins, in a series of two-electron transfer reactions (56). The nitroimidiazole drug metronidazole (MTZ) is a critical component of important combination therapies against *H. pylori*. However, resistance to at least moderate levels of MTZ is very common, ranging up to 90% of isolates in some developing countries, and constitutes a major cause of treatment failures (57). The development of moderate-level MTZ resistance in *H. pylori* generally depends on mutational inactivation of the *rdx*A (NADPH nitroreductase) gene (HP0954), with higher level resistance resulting from loss of function mutations in additional reductase genes (58–60). Loss of metabolic enzymes, even apparently dispensable ones such as these nitroreductases, will often be compensated by changes in other metabolic functions. Previous physiologic tests suggest that exposure of genetically resistant *H. pylori* strains to sublethal concentrations of MTZ results in changes in levels of some key enzymes, which could enable these strains realize their resistance potential (61). Proteome technologies may be used to examine the protein profile of a MTZ^R derivative of reference strain 26695 grown in medium with a moderate amount of MTZ. The proteins from lysed cell pellets of *H. pylori* MTZ^R grown in the absence and presence of MTZ were separated on 2D gels run in parallel using an initial pH gradient of pH 4 to pH 9. The Sypro stained gel (see Figure 5) of the sample grown in the absence of MTZ revealed many prominent individual protein spots. Forty-two individual protein spots were chosen which showed differential regulation between the two gels. Two-dimensional spots from the pH 4–10 gel were analyzed by *in situ* digestion with Lys-C followed by MALDI-TOF analysis. The identifications of these spots have been described (62). Of the gel spots identified from the *H. pylori* genomic database, spots 11 and 12, spots 19 and 20, spot 22, and spots 29–42 consistently showed differential expression when *H. pylori* was grown in the presence of metronidazole. Fluorescent intensities of spots 11 and 12 (isocitrate dehydrogenase) appeared to drop by approximately 50% when cells were grown in the presence of metronidazole. Likewise, fluorescence intensities in spots 29–42 appeared to drop by at least twofold when the bacteria was grown in the presence of metronidazole. However, the most obvious difference in expression was observed in spots 19, 20, and 22, which appeared to be upregulated when the MTZ-resistant strain of *H. pylori* was grown in the presence of metronidazole. Spots 19 and 20 were identified as alkylhydroperoxide reductase (*ahp*C). Spot 22 corresponded to aconitase B. As to be expected, a considerable number of proteins from various metabolic pathways were repressed by the addition of MTZ to the growth media. The proteomic approach provided a broader screen, and allowed the detection in particular, of a compensatory increase in expression of a set of proteins encoded by HP1563, alkyl hydroperoxide reductase (*ahp*C), which has not been reported in *H. pylori*. The question was this: why *ahp*C?

Incomplete reduction of molecular oxygen leads to the formation of reactive oxygen species that can damage nearly all subcellular macromolecules in some form or another (63). Cells are at least partially protected against these reactive oxygen species by superoxide dismutase, which eliminates the O_2^- radical, and catalase, which removes H_2O_2, and peroxidases, which remove alkyl peroxides as well as H_2O_2 (64). Alkyl hydroperoxidase enzymes reduce hydroperoxide intermediates with electrons donated by NADPH via thioredoxin or other thiol-

Mtz^r *Helicobacter pylori* pH 3-10 proteomic map

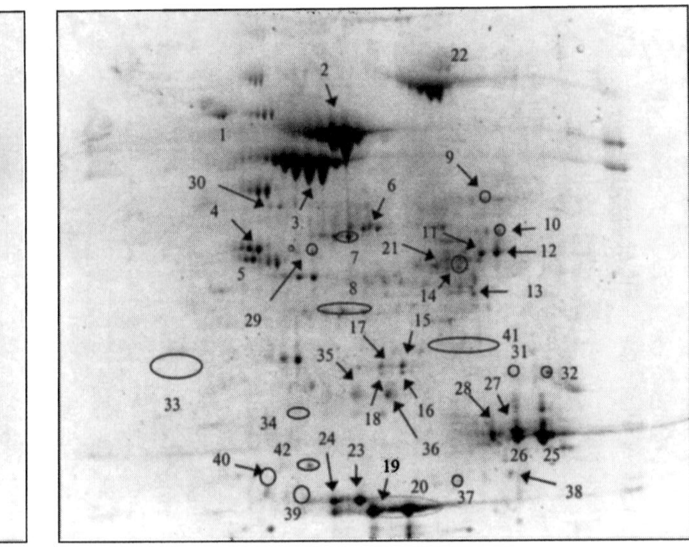

Metronidazole Resistant
Helicobacter pylori

Metronidazole Resistant
Helicobacter pylori
+ metronidazole

Figure 5 *Helicobacter pylori* two-dimensional (2D) map (pH gradient electrophoresis pH 4–10) with names of identified proteins (listed in Table I). *H. pylori* MET^R was grown as described in Materials and Methods. 100 μg of protein extract was loaded in the first dimension. (A) Sypro Ruby Red stained 2D gel of *H. pylori* MET^R grown in the absence of metronidazole. (B) Sypro Ruby Red stained 2D gel of *H. pylori* MET^R grown in the presence of 18 μg/ml of metronidazole. (Figure reprinted with permission from Wiley–VHC)

containing intermediates, and the alkyl hydroperoxidase reductase family seen in *H. pylori* represents a new type of peroxidase that has a conserved cysteine as the primary site of catalysis rather than the selenocysteine of glutathione peroxidase (65). Antioxidant defense mechanisms are induced as part of a global cellular adaptive response to oxidative stress in many organisms (66). In previous studies evaluating clinical isolates of isoniazid resistant *Mycobacterium tuberculosis* defective in *kat*G, a conserved essential gene, it has been shown that isoniazid-resistant *kat*G mutants compensate for the loss of *kat*G activity by a second mutation, resulting in hyperexpression of at least one isoform of *ahp*C (67). *Ahp*C in *H. pylori* has been found to upregulated in response to iron deprivation (68). If this gene is knocked out, the organism will suffer from oxygen toxicity problems (69). Two possibilities might explain why there appears to be differential expression of isoforms of *ahp*C: The obvious answer would be phosphorylation; the other, dissociation of an iron–sulfur center similar to *oxy*R. In other words, *ahp*C may be a redox-responsive protein and in that respect may interact with *rdx*A. Regardless, alkyl hydroperoxidase identification could lead to a worthwhile drug target, identified via proteomic analysis.

REAL-WORLD APPLICATION 3: IDENTIFICATION OF SURROGATE MARKERS FOR BACTERIAL CELL DIVISION AND POTENTIAL CELL-BASED SCREENING ASSAYS

An area that has been largely unexploited as a potential source for novel antibacterial targets is the complex mechanism surrounding bacterial cell division. This is due largely to the difficulty of devising a high-throughput screen for cell-division inhibitors. Bacterial cell division is dependent on a set of key division proteins that are conserved across a broad range of bacteria and differs significantly from the eukaryotic division apparatus. These proteins were initially identified in *E. coli* as temperature-sensitive mutations that led to filamentous growth (i.e., *f*ilamenting *t*emperature *s*ensitive, leading to very long, nonseptate cells) (70). Two of these proteins (e.g., ftsA and ftsZ) are essential to the process of division and have been shown to participate either directly or indirectly in the assembly of a cell-division apparatus (71,72). These proteins are localized with other division-associated proteins, at the time of initiation of cell division, to a ringlike structure at the future cell-division site. The most extensively studied bacterial cell-division protein is the prototubulin gene product ftsZ. This protein, conserved among a very broad range of bacteria, is the earliest protein to assemble at the division site. It has been found to have GTPase activity and has been demonstrated to form GTP-dependent polymers *in vitro* (73). This *in vitro* assembly mimics the *in vivo* assembly of the ring structure at the bacterial cell-division site. The actin like protein ftsA, also present in all known eubacteria, plays an essential role in septation and cell division (74). FtsA behaves as a dimer and maintains both ATP binding and ATP hydrolysis activities (75).

The use of proteomics is proposed to determine whether the inhibition of cell division in bacteria leads to cell-cycle-specific changes that can be used to construct an indicator strain to screen for division inhibitors. It is important to note that any changes in the proteome using temperature-sensitive *fts*A or *fts*Z mutants must differentiate between changes that are specific for division versus those that are due to general stress responses or the shift to restrictive temperature. The proteome of a filamenting temperature-sensitive mutant (*fts*A) of *E. coli* that was unable to divide into daughter cells during growth at 42°C was examined. Comparison of 2D electrophoresis profiles generated from cell lysates of *fts*A and the wild-type isogenic strain grown at the restricted temperature (42°C) allowed for the identification of protein spots that were specifically up- or downregulated in the mutant strain. These spots were subjected to in-gel proteolytic digestion followed by mass spectrometry analysis for peptide sequence and identity. From these results, several gene products that were differentially expressed between the two bacterial phenotypes were identified (filamented versus normal morphology) (see Figure 6). In this experimental setting, *E. coli* strain MC4100 and an isogenic *fts*A temperature-sensitive cell-division strain, constructed via P1 phage transduction, were utilized. The shift to restrictive temperature led to the loss of activity in FtsA resulting in filamentous growth of the cells. Cell lysates were compared at both permissive (30°C) and restrictive (42°C) temperatures (see Figure 7), and several gene products from the MC4100 and *fts*A strains were differentially expressed. In this gel profile, uridine phosphorylase, phosphoglycerate mutase I, and periplasmic ribose binding protein are shown (right to left in the figure). In this case, it appears that uridine phosphorylase represents a potential candidate for the proposed surrogate marker assay, whereas the other proteins likely do not. Another example is shown in Figure 8, in which a typical surrogate marker response is compared to another cell septation marker, F1 ATP synthase (76).

From this information, Lac reporter fusion vectors can be constructed to test these surrogate marker proteins for utility in a high-throughput cell-based screening assay. From such studies, it may be possible to identify division-specific markers, that could be used in cell-based assays to screen for novel small-molecule inhibitors of the cell-division pathway. Such inhibitors could be developed into potentially novel antibiotics, targeting a pathway specific for and conserved among gram-positive and gram-negative bacterial species.

In a similar manner, known antimicrobial compounds or temperature-sensitive strains in potential targets could be utilized to generate initial proteomic data, thereby establishing guideposts useful for uncovering new structural classes of compounds. This platform could potentially be suitable for evaluation and discovery of other inhibitors of cell processes in both gram-negative and gram-positive

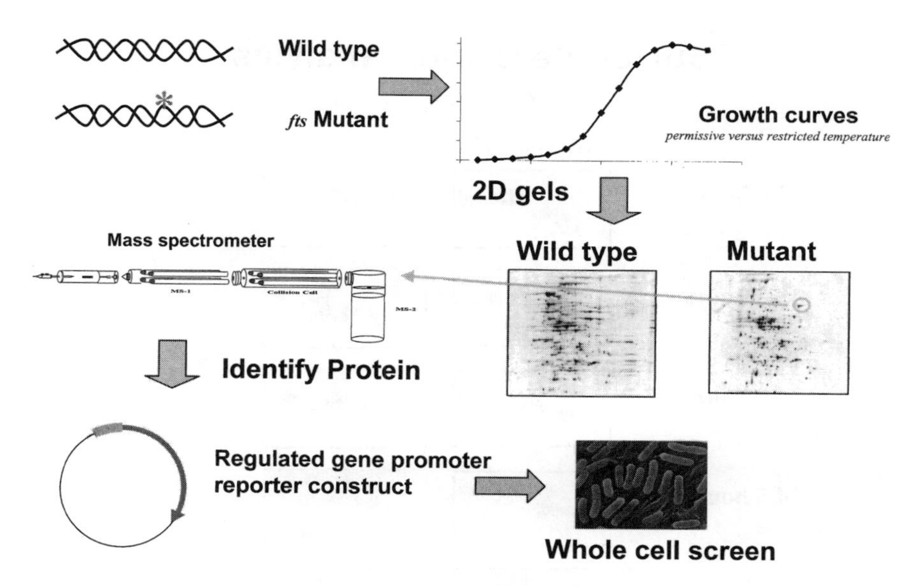

FIGURE 6 Cell based surrogate marker strategy. An *fts* mutant and a wt strain are grown at both permissive and restricted temperatures. Cell lysates are then evaluated on 2D gels and differentially expressed spots are analyzed. Once the potential surrogate marker for cell division is identified a reporter gene construct is generated followed by whole cell screening.

bacteria. There have been many published examples of proteomically derived antimicrobial targets for therapeutic intervention such as *Pseudomonas aeruginosa* from chronic lung infections in cystic fibrosis (CF) patients, *Staphlococcus aureus*, *M. tuberculosis*, *Francisella tularensis*, *Haemophilus influenzae*, *Candida albicans*, and *Salmonella* (77–84). It has lately been heartening to see biology and medicine finally starting to take hold in (or, rather, "infect") this new technology. It is important, however, to recognize that as with any technology, there are limitations to proteomic approaches. This point is often lost or obscured by articles touting the "wonders of technology." Therefore, it is critical to note that while proteomics offers a unique approach to the global analysis of cellular proteins, validation of the results can *only* be achieved through studies based on biological phenomena in conjunction with complementary approaches such as molecular biology. The development of antimicrobial therapies mesh well with the intended goals of modern proteome technologies as there is a plethora of readily available bacterial strains and mutants from various key metabolic pathways to study with relatively little constraint from lack of cost-effective reagents. Although previous

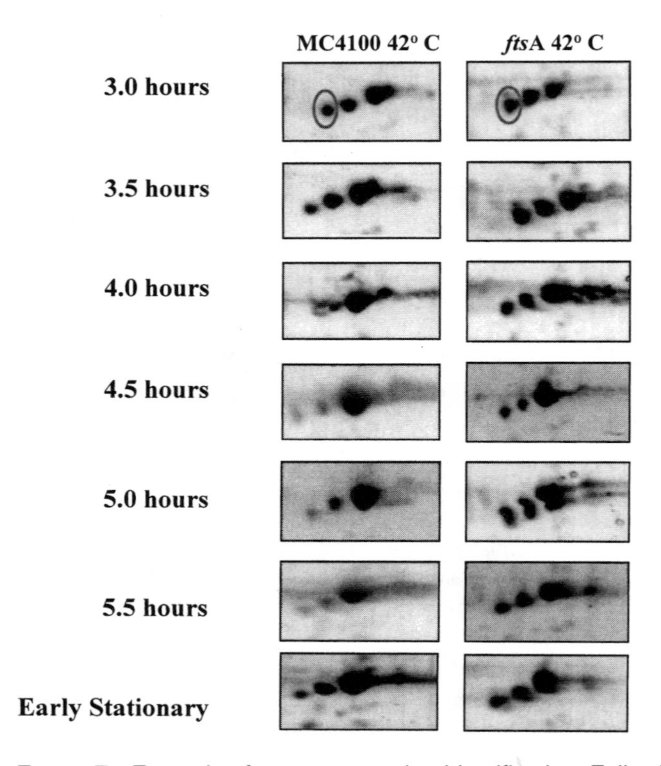

FIGURE 7 Example of surrogate marker identification. Following 2D analysis of wt versus *ftsA E. coli*, a cluster of three proteins was identified; (from left to right) Uridine phosphorylase, phosphoglycerate mutase I, and periplasmic ribose binding protein. Uridine phosphorylase is circled in the upper panel. From the information and mean fluorescence derived from these Sypro Red stained spots, uridine phosphorylase appears to be coordinately regulated with the cell septation process and represents a potential surrgate marker.

biochemical characterization of these mutants has been very challenging and identification of many of their regulatory components and functions remains tricky, many of their genome sequences have been recently deciphered, so the lexicon is in place to enable the proteomic approach. Through proteome technologies, it should be possible in future experimentation to more fully address many

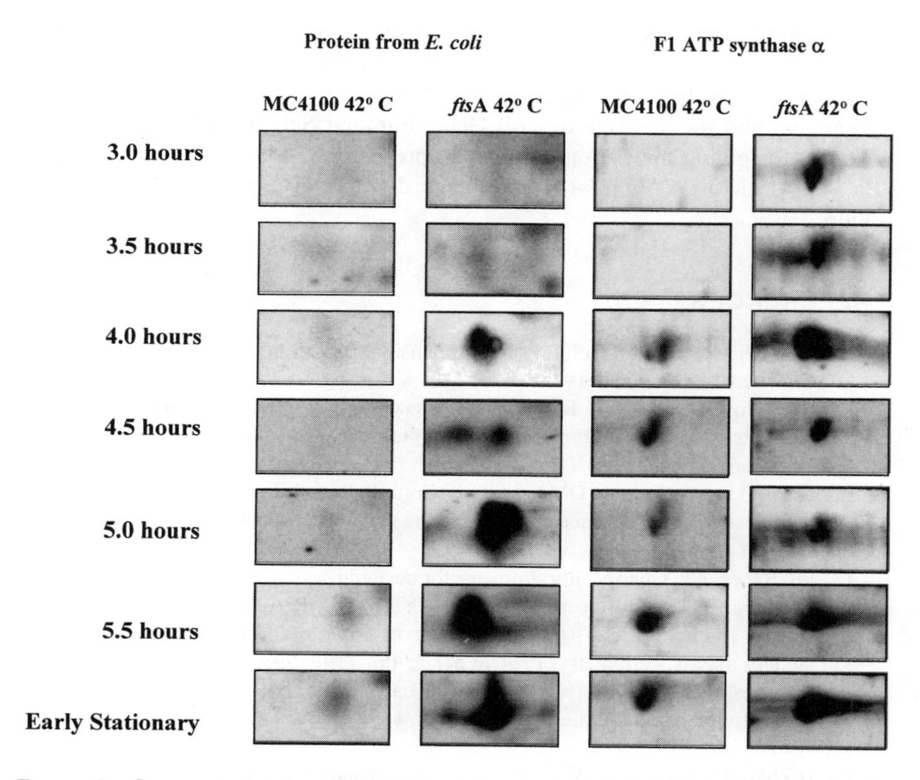

Figure 8 Surrogate markers for a filamenting mutant. In another 2D analysis of wt versus an *fts* mutant, one can clearly see the evolution of a potential surrogate marker from the Sypro staining pattern. This is compared to F1 ATP synthase α which is known to be a marker for active cell division.

of the central issues of bacterial growth, metabolism, and resistance that have gone unanswered. However, let's get real! Proteomics won't do this single-handedly.

ACKNOWLEDGMENTS

I thank two of my long time *H. pylori* collaborators, Drs. Doug Berg of Washington University School of Medicine and Paul Hoffman of Dalhousie University

Faculty of Medicine, who have been very supportive of this type of work (biology directed proteomics) before proteomics became a household word. Other people who deserve mention are the "Three Horsemen" of antimicrobial drug development, formerly of Bristol-Myers Squibb, now at Pfizer, Achillion, and Merck: Tom Dougherty, Mike Pucci, and John Barrett, who have been loyal supporters without reserve. John Barrett is the template for what a great corporate manager should be. Last, but certainly not least, I'd like to thank my daughters, Erin and Shannon, and the students of Toll Gate Grammar School in Pennington, New Jersey, for giving me the opportunity to present the wonder of Louis Pasteur á la Paul de Kruif's *Microbe Hunters*. We don't need more scientists, we need more Pasteurs!

REFERENCES

1. PH O'Farrell. High resolution two-dimensional electrophoresis of proteins. *J Biol Chem 250*:4007–4021, 1975.
2. B Bjellqvist, K Ek, PG Righetti, E Gianazza, A Gorg, R Westermeier, W Postel. Isoelectric focusing in immobilized pH gradients: principle, methodology and some applications. *J Biochem Biophys Methods 6*:317–339, 1982.
3. DN Perkins, DJ Pappin, DM Creasy, JS Cottrell. Probability-based protein identification by searching sequence databases using mass spectrometry data. *Electrophoresis 20*:3551–3567, 1999.
4. PR Baker, KR Clauser. http://prospector.ucsf.edu.
5. A Ducret, I Van Oostveen, JK Eng, JR Yates 3rd, R Aebersold. High throughput protein characterization by automated reverse-phase chromatography/electrospray tandem mass spectrometry. *Protein Sci 7*:706–719, 1998.
6. MP Washburn, D Wolters, JR Yates 3rd. Large-scale analysis of the yeast proteome by multidimensional protein identification technology. *Nat Biotechnol 19*:242–247, 2001.
7. H Zhou, JA Ranish, JD Watts, R Aebersold. Quantitative proteome analysis by solid-phase isotope tagging and mass spectrometry. *Nat Biotechnol 20*:512–515, 2002.
8. SP Gygi, B Rist, SA Gerber, F Turecek, MH Gelb, R Aebersold. Quantitative analysis of complex protein mixtures using isotope-coded affinity tags. *Nat Biotechnol 17*: 994–999, 1999.
9. BB Haab. Advances in protein microarray technology for protein expression and interaction profiling. *Curr Opin Drug Discov Devel 4*:116–123, 2001.
10. D Figeys, D Pinto. Proteomics on a chip: promising developments. *Electrophoresis 22*:208–216, 2001.
11. M Eck, B Schmausser, R Haas, A Greiner, S Czub, H Muller-Hermelink. MALT-type lymphoma of the stomach is associated with *Helicobacter pylori* strains expressing the CagA protein. *Gastroenterology 112*:1482–1486, 1997.
12. SJ Riegg, BE Dunn, M Blaser. Microbiology and pathogenesis of *Helicobacter pylori*. In: MJ Blaser, PD Smith, JI Ravdin, HB Greenberg, and RL Guerrant, eds. *Infections of the gastrointestinal tract*. New York: Raven Press, 1995, pp. 535–550.

13. A Lee. Animal models and vaccine development. *Bailliere's Clin Gastroenterol 9*: 615–632, 1995.
14. CK Lee, R Weltzin, WD Thomas Jr, H Kleanthous, TH Ermak, G Soman, JE Hill, SK Ackerman, TP Monath. Oral immunization with recombinant *Helicobacter pylori* urease induces secretory IgA antibodies and protects mice from challenge with *Helicobacter felis*. *J Infect Dis 172*:161–172, 1995.
15. M Kostrzynska, JD Betts, JW Austin, TJ Trust. Identification, characterization, and spatial localization of two flagellin species in *Helicobacter pylori* flagella. *J Bacteriol 173*:937–946, 1991.
16. DY Graham, MF Go, DJ Evans Jr. Urease, gastric ammonium/ammonia, and *Helicobacter pylori*—the past, the present, and recommendations for future research. *Aliment Pharmacol Ther 6*:659–669, 1992.
17. TL Cover, MJ Blaser. Purification and characterization of the vacuolating toxin from *Helicobacter pylori*. *J Biol Chem 267*:10570–10575, 1992.
18. D Ilver, A Arnqvist, J Ögren, I-M Frick, D Kersulyte, ET Incecik, DE Berg, A Covacci, L Engstrand, T Boren. *Helicobacter pylori* adhesin binding fucosylated histo-blood group antigens revealed by retagging. *Science 279*:373–377, 1998.
19. S Censini, C Lange, Z Xiang, JE Crabtree, P Ghiara, M Borodovsky, R Rappuoli, A Covacci. cag, a pathogenicity island of *Helicobacter pylori*, encodes type I-specific and disease-associated virulence factors. *Proc Natl Acad Sci USA 93*:14648–14653, 1996.
20. NS Akopyants, SW Clifton, D Kersulyte, JE Crabtree, BE Youree, CA Reece, NO Bukanov, ES Drazek, BA Roe, DE Berg. Analyses of the *cag* pathogenicity island of *Helicobacter pylori*. *Mol Microbiol 28*:37–53, 1998.
21. B Kimmel, A Bosserhoff, R Frank, R Gross, W Goebel, D Beier. Identification of immunodominant antigens from Helicobacter pylori and evaluation of their reactivities with sera from patients with different gastroduodenal pathologies *Infect Immun 68*:915–920, 2000.
22. S Censini, C Lange, Z Xiang, JE Crabtree, P Ghiara, M Borodovsky, R Rappuoli, A Covacci. *cag*, a pathogenicity island of *Helicobacter pylori*, encodes type I-specific and disease-associated virulence factors. *Proc Natl Acad Sci USA 93*:14648–14653, 1996.
23. J-F Tomb et al. The complete genome sequence of the gastric pathogen *Helicobacter pylori*. *Nature 388*:539–547, 1997.
24. CP McAtee, KE Fry, DE Berg. Identification of potential diagnostic and vaccine candidates of Helicobacter pylori by 'proteome' technologies. *Helicobacter 3*: 163–169, 1998b.
25. MJM Buckley, M Deltenre. Therapy of *H. pylori* infection. *Curr Opin Gastroenterol 13*:56–62, 1997.
26. RL Ferrero, J-M Thiberge, I Kansau, N Wuscher, M Huerre, A Labigne. The GroES homolog of *Helicobacter pylori* confers protective immunity against mucosal infection in mice. *Proc Natl Acad Sci USA 92*:6499–6503, 1995.
27. M Marchetti, M Rossi, V Gianelli, MM Giuliani, M Pizza, S Censini, A Covacci, P Massari, C Pagliaccia, R Manetti, JL Telford, G Douce, G Dougan, R Rappuoli, P Ghiara. Protection against *Helicobacter pylori* infection in mice by intragastric

vaccination with *H. pylori* antigens is achieved using a non-toxic mutant of *E. coli* heat-labile enterotoxin (LT) as adjuvant. *Vaccine 16*:33–37, 1998.

28. J Pappo, WD Thomas Jr, Z Kabok, N Taylor, JC Murphy, JG Fox. Effect of oral immunization with recombinant urease on murine *Helicobacter felis* gastritis. *Infect Immun 63*:1246–1252, 1995.
29. FJ Radcliff, SL Hazell, T Kolesnikow, C Doidge, A Lee. Catalase, a novel antigen for *Helicobacter pylori* vaccination. *Infect Immun 65*:4668–4674, 1997.
30. OG Gomez-Duarte, B Lucas, Z-X Yan, K Panthel, R Haas, TF Meyer. Protection of mice against gastric colonization by *Helicobacter pylori* by single oral dose immunization with attenuated *Salmonella typhimurium* producing urease subunits A and B. Vaccine *16*:460–471, 1998.
31. P Ghiara, M Rossi, M Marchetti, A Di Tommaso, C Vindigni, F Ciampolini, A Covacci, JL Telford, MT De Magistris, M Pizza, R Rappuoli, G Del Giudice. Therapeutic intragastric vaccination against *Helicobacter pylori* in mice eradicates an otherwise chronic infection and confers protection against reinfection. *Infect Immun 65*:4996–5002, 1997.
32. CP McAtee, MY Lim, K Fung, M Velligan, K Fry, T Chow, DE Berg. Identification of potential diagnostic and vaccine candidates of *Helicobacter pylori* by two-dimensional gel electrophoresis, sequence analysis, and serum profiling. *Clin Diagn Lab Immunol 5*:537–542, 1998.
33. BE Dunn, GP Campbell, GI Perez-Perez, MJ Blaser. Purification and characterization of urease from *Helicobacter pylori. J Biol Chem 265*:9464–9469, 1990.
34. BE Dunn, R Roop, C-C Sung, S Sharma, GI Perez-Perez, MJ Blaser. Identification and purification of a cpn60 heat shock protein homolog from *Helicobacter pylori. Infect Immun 60*:1946–1951, 1992.
35. NJ Hughes, PA Chalk, CL Clayton, DJ Kelly. Identification of carboxylation enzymes and characterization of a novel four-subunit pyruvate:flavodoxin oxidoreductase from *Helicobacter pylori. J Bacteriol* 177:3953–3959, 1995.
36. W Schmitt, S Odenbreit, D Heuermann, R Haas. Cloning of the *Helicobacter pylori* recA gene and functional characterization of its product. *Mol Gen Genet 248*: 563–572, 1995.
37. JF Tomb et al., 1997 *ibid.*
38. M Kostrzynska, JD Betts, JW Austin, TJ Trust. Identification, characterization, and spatial localization of two flagellin species in *Helicobacter pylori* flagella. *J Bacteriol 173*:937–946, 1991.
39. KM Ottemann, AC Lowenthal. *Helicobacter pylori* uses motility for initial colonization and to attain robust infection. *Infect Immun 70*:1984–1990, 2002.
40. WS Ji, JL Hu, JW Qiu, DR Peng, BL Shi, SJ Zhou, KC Wu, DM Fan. Polymorphism of flagellin A gene in *Helicobacter pylori. World J Gastroenterol 6*:783–787, 2001.
41. Y Fukui, I Saito, K Shiroki, H Shimojo, Y Takebe, Y Kaziro. The 19-kDa1 protein encoded by early region 1b of adenovirus type 12 is synthesized efficiently in *Escherichia coli* only as a fused protein. *Gene 23*:1–13, 1983.
42. E Kasimoglu, SJ Park, J Malek, CP Tseng, RP Gunsalus. Transcriptional regulation of the proton-translocating ATPase (*atpIBEFHAGDC*) operon of *Escherichia coli*: control by cell growth rate. *J Bacteriol 178*:5563–5567, 1996.

43. Q Strolin, M Benedetti, J Thomassin, P Tocchetti, P Dostert, R Kettler, M Da Prada. Species differences in changes of heart monoamine oxidase activities with age. *J Neural Transm* 1994; 41(suppl):83–87.
44. DJ Evans Jr, DG Evans, HC Lampert, H Nakano. Identification of four new prokaryotic bacterioferritins, from *Helicobacter pylori, Anabaena variabilis, Bacillus subtilis* and *Treponema pallidum*, by analysis of gene sequences. *Gene 153*:123–127, 1995.
45. JF Tomb et al., 1997. ibid
46. D Missiakas, F Schwager, JM Betton, C Georgopoulos, S Raina. Identification and characterization of HsIV HsIU (ClpQ ClpY) proteins involved in overall proteolysis of misfolded proteins in *Escherichia coli. EMBO J 15*:6899–6909, 1996.
47. JF Tomb et al., 1997. ibid
48. J Velayudhan, NJ Hughes, AA McColm, J Bagshaw, CL Clayton, SC Andrews, DJ Kelly. Iron acquisition and virulence in Helicobacter pylori: a major role for FeoB, a high-affinity ferrous iron transporter. *Mol Microbiol 37*:274–286, 2000.
49. PW O'Toole, SM Logan, M Kostrzynska, T Wadstrom, TJ Trust. Isolation and biochemical and molecular analyses of a species-specific protein antigen from the gastric pathogen *Helicobacter pylori. J Bacteriol 173*:505–513, 1991.
50. SM Pitson, GL Mendz, S Srinivasan, SL Hazell. The tricarboxylic acid cycle of *Helicobacter pylori. Eur J Biochem 260*:258–267, 1999.
51. JF Tomb et al. 1997. ibid
52. SL Hazell, DJ Evans Jr, DY Graham. *Helicobacter pylori* catalase. *J Gen Microbiol 137*:57–61, 1991.
53. DJ Evans Jr, DG Evans. Helicobacter pylori adhesins: review and perspectives. *Helicobacter 5*:183–195, 2000.
54. MM Exner, P Doig, TJ Trust, RE Hancock. Isolation and characterization of a family of porin proteins from *Helicobacter pylori. Infect Immun 63*:1567–1572, 1995.
55. BE Dunn, GP Campbell, GI Perez-Perez, MJ Blaser. Purification and characterization of urease from *Helicobacter pylori. J Biol Chem 265*:9464–9469, 1990.
56. NR Krieg, PS Hoffman. Microaerophily and oxygen toxicity. *Annu Rev Microbiol 40*:107–130, 1986.
57. GD Bell, K Powell, SM Burridge, A Pallecaros, PH Jones, PW Gant, G Harrison, JE Trowell. Experience with "triple" anti-*Helicobacter pylori* eradication therapy: side effects and the importance of testing the pre-treatment bacterial isolate for metronidazole resistance. *Aliment Pharmacol Ther 6*:427–435, 1992.
58. A Goodwin, D Kersulyte, G Sisson, SJO Veldhuyzen van Zanten, DE Berg, PS Hoffman. Metronidazole resistance in *Helicobacter pylori* is due to null mutations in a gene (*rdxA*) that encodes an oxygen-insensitive NADPH nitroreductase. *Mol Microbiol 28*:383–393, 1998.
59. PJ Jenks, RL Ferrero, A Labigne. The role of the *rdxA* gene in the evolution of metronidazole resistance in *Helicobacter pylori. J Antimicrob Chemother 43*: 753–758, 1999a.
60. JY Jeong, AK Mukhopadhyay, D Dailidiene, Y Wang, B Velapatiño, RH Gilman, AJ Parkinson, GB Nair, BCY Wong, SK Lam, R Mistry, I Segal, Y Yuan, H Gao, T Alarcon, ML Brea, Y Ito, D Kersulyte, H-K Lee, Y Gong, A Goodwin, PS Hoffman, DE Berg. Sequential inactivation of *rdxA* (HP0954) and *frxA* (HP0642) nitroreductase

genes cause moderate and high-level metronidazole resistance in *Helicobacter pylori*. *J Bacteriol 182*:5082–5090, 2000.
61. PS Hoffman, A Goodwin, J Johnsen, K Magee, SJ Veldhuyzen van Zanten. Metabolic activities of metronidazole-sensitive and -resistant strains of *Helicobacter pylori*: repression of pyruvate oxidorductase and expression of isocitrate lyase activity correlate with resistance. *J Bacteriol 178*:4822–4829, 1996.
62. CP McAtee, PS Hoffman, DE Berg. Identification of differentially regulated proteins in metronidozole resistant *Helicobacter pylori* by proteome techniques. *Proteomics 1*:516–521, 2001.
63. H Sies. Strategies of antioxidant defense. *Eur J Biochem 215*:213–219, 1993.
64. G Storz, FS Jacobson, LA Tartaglia, RW Morgan, LA Silveira, BN Ames. An alkyl hydroperoxide reductase induced by oxidative stress in *Salmonella typhimurium* and *Escherichia coli*: genetic characterization and cloning of *ahp*. *J Bacteriol 171*: 2049–2055, 1989.
65. MK Cha, CH Yun, IH Kim. Interaction of human thiol-specific antioxidant protein 1 with erythrocyte plasma membrane. *Biochemistry 39*:6944–6950, 2000.
66. LM Baker, A Raudonikiene, PS Hoffman, LB Poole. Essential thioredoxin-dependent peroxiredoxin system from *Helicobacter pylori*: genetic and kinetic characterization. *J Bacteriol 183*:1961–73, 2001.
67. DR Sherman, K Mdluli, MJ Hickey, TM Arain, SL Morris, CE Barry 3rd, CK Stover. Compensatory ahpC gene expression in isoniazid-resistant *Mycobacterium tuberculosis*. *Science 272*:1641–1643, 1996.
68. ML Baillon, AH van Vliet, JM Ketley, C Constantinidou, CW Penn. An iron-regulated alkyl hydroperoxide reductase (AhpC) confers aerotolerance and oxidative stress resistance to the microaerophilic pathogen *Campylobacter jejuni*. *J Bacteriol 181*:4798–804, 1999.
69. SN Wai, K Nakayama, K Umene, T Moriya, K Amako. An iron-regulated alkyl hydroperoxide reductase (ahpC) confers aerotolerance and oxidative stress resistance to the microaerophilic pathogen *Campylobacter jejuni*. *J Bacteriol 181*:4798–4804, 1999.
70. KJ Begg, WD Donachie. Cell shape and division in *Escherichia coli*: experiments with shape and division mutants. *J Bacteriol 163*:615–622, 1985.
71. SG Addinall, E Bi, J Lutkenhaus. FtsZ ring formation in *fts* mutants. *J Bacteriol 178*:3877–3884, 1996.
72. A Descoteaux, GR Drapeau. Regulation of cell division in *Escherichia coli* K-12: probable interactions among proteins FtsQ, FtsA, and FtsZ. *J Bacteriol 169*: 1938–1942, 1987.
73. E Nogales, KH Downing, LA Amos, J Lowe. Tubulin and FtsZ form a distinct family of GTPases. *Nat Struct Biol 5*:451–458, 1998.
74. QM Yi, S Rockenbach, JE Ward Jr, J Lutkenhaus. Structure and expression of the cell division genes *fts*Q, *fts*A and *fts*Z. *J Mol Biol 184*:399–412, 1985.
75. P Bork, C Sander, A Valencia. An ATPase domain common to prokaryotic cell cycle proteins, sugar kinases, actin, and hsp70 heat shock proteins. *Proc Natl Acad Sci USA 89*:7290–7294, 1992.
76. W Junge, O Panke, DA Cherepanov, K Gumbiowski, M Muller, S Engelbrecht. Inter-subunit rotation and elastic power transmission in F0F1-ATPase. *FEBS Lett 504*:152–160, 2001.

77. CL Nilsson. Bacterial proteomics and vaccine development. *Am J Pharmacogenom 2*:59–65, 2002.
78. SJ Cordwell, AS Nouwens, BJ Walsh. Comparative proteomics of bacterial pathogens. *Proteomics 1*:461–72, 2001.
79. SL Hanna, NE Sherman, MT Kinter, JB Goldberg. Comparison of proteins expressed by *Pseudomonas aeruginosa* strains representing initial and chronic isolates from a cystic fibrosis patient: an analysis by 2-D gel electrophoresis and capillary column liquid chromatography-tandem mass spectrometry. *Microbiology 146*:2495–508, 2000.
80. MG Sonnenberg, JT Belisle. Definition of *Mycobacterium tuberculosis* culture filtrate proteins by two-dimensional polyacrylamide gel electrophoresis, N-terminal amino acid sequencing, and electrospray mass spectrometry. *Infect Immun 65*: 4515–4524, 1997.
81. L Hernychova, J Stulik, P Halada, A Macela, M Kroca, T Johansson, M Malina. Construction of a *Francisella tularensis* two-dimensional electrophoresis protein database. *Proteomics 1*:508–515, 2001.
82. DN Chakravarti, MJ Fiske, LD Fletcher, RJ Zagursky. Application of genomics and proteomics for identification of bacterial gene products as potential vaccine candidates. *Vaccine 19*:601–612, 2000.
83. A Pitarch, R Diez-Orejas, G Molero, M Pardo, M Sanchez, C Gil, C Nombela. Analysis of the serologic response to systemic *Candida albicans* infection in a murine model. *Proteomics 1*:550–559, 2001.
84. J Deiwick, C Rappl, S Stender, PR Jungblut, M Hensel. Proteomic approaches to *Salmonella* Pathogenicity Island 2 encoded proteins and the SsrAB regulon. *Proteomics 2*:792–799, 2002.

11

Surrogate Ligand-Based Assay Systems for Discovery of Antibacterial Agents for Genomic Targets

Dale J. Christensen and Paul T. Hamilton
Karo Bio USA, Inc., Durham, North Carolina, U.S.A.

INTRODUCTION

Since the introduction of penicillin and sulfa antibiotics, numerous antibacterial agents have been produced to treat infectious diseases, leading to dramatic reductions in illness and death. By the 1980s, it was believed that industrialized nations had developed all of the tools necessary to control microbial pathogens (1). However, widespread use of antibiotics provided powerful selective pressure for development of mechanisms that create resistance to antibacterial agents. These resistance mechanisms have spread through bacterial populations so pervasively that antibiotic-resistant strains seriously compromise the successful treatment of infectious diseases. Thus, discovery and development of new antibiotics that act on novel targets are critical to the ability to treat bacterial infections in the future. Knowledge of the biology of individual pathogens, in conjunction with genomic sequence information, may lead to the development of highly specific antibiotics. Genomic information could also be used to identify common targets in many species of bacteria for the design of broad-spectrum antibiotics.

During the past several years, the development of technology for efficient genomic sequencing has resulted in the elucidation of complete DNA sequences nearly 40 bacterial genomes, and the sequencing of another 100 bacterial genomes is in progress (2–7). Since many of these genomes are from human pathogens, a wide range of targets involved in diverse microbial pathways is now accessible for antibacterial drug discovery. The basic criterion for selection of a gene product as an antibacterial target is that it be essential for the survival of the pathogen in the host (8). Unfortunately, a significant number of the genes determined to be essential do not have a known biological function (1–3) and most current target-based screening strategies rely on biochemical activity to screen for inhibitors. The *Escherichia coli* genome sequence, for example, contains genes with no known function for approximately 40% of the identified genes (3). Many of these genes of unknown function are essential for cell growth, but a large number of these targets will be impractical to screen using conventional HTS methods (8). Thus, there is a clear need for technologies that allow for screening of previously unscreenable targets to accelerate the antibacterial drug discovery process. Two main approaches have been developed to fill the void of function-blind screening, direct binding assays and surrogate ligand competition assays.

Direct binding of a compound to a target protein can be measured by several methods and has been exploited for the development of assays for targets of unknown function. When a compound binds to a protein, the thermal melting temperature of the protein changes and can be detected by microcalorimetry or binding of fluorescent probes (9,10). Specialized equipment has been developed by Scriptgen and 3D-Pharmaceuticals to perform these assays through thermal melting. However, compounds detected in these assays may bind at any site on the target protein, including sites that are not critical for the function of the target protein or sites that are only exposed during the thermal cycling. Therefore, these assays cannot be validated and the compounds identified from these assays must be individually validated.

Another method for direct binding assays, developed by Cetek, uses capillary electrophoresis to identify compounds that bind to the target protein (11). The target protein is incubated with compounds and changes in the retention time of the protein during capillary electrophoresis are used to detect binding of compounds to the target. This method offers an advantage over other direct binding assays by eliminating thermal cycling, which partially denatures the protein, so compound binding occurs only at sites on the target that exist in the native conformation. However, the binding site of the compound must be independently validated for inhibition of the target function.

Surrogate ligand competition assays represent another technology to assay targets of unknown function. These assays rely on the use of peptides, or other molecular probes, as surrogate ligands in a competitive binding assay (12,13). This techniques has an advantage because compounds are detected only when they bind at the binding site of the probe. Therefore, the assay can be validated prior to high-throughput screening by demonstration that the probe binds at a site that is critical for the function of the target (14,15). The remainder of this chapter summarizes the techniques available for identification, validation, and utilization of surrogate ligand probes to identify novel small molecule drug leads.

PEPTIDE IDENTIFICATION TECHNOLOGIES

Peptide display technologies are based on a physical linkage between phenotype, the peptide, and genotype and the nucleic acid encoding the peptide. One of the most widely practiced methods for identifying peptides that specifically bind to a target protein is phage display. George Smith first described phage display (16,17) in 1985 when random DNA fragments were inserted into gene III of a filamentous bacteriophage and the polypeptides were displayed on the surface of the bacteriophage as fusions to the pIII coat protein. These "fusion phage" could then be selected over ordinary phage based on affinity purification with an antibody directed against the cloned polypeptide. After this initial demonstration of phage display technology, random peptide libraries were generated in phage display systems and used for epitope mapping of antibodies or to identify mimetic peptides (18–20). Phage display of peptide libraries has since proven to be a powerful tool that can be used to isolate surrogate ligands for drug discovery (12,21) and affinity chromatography (22,23), to study protein–protein interactions (24), and to identify peptides that are directed to organs or tissues (25,26).

Originally, phage display was developed with peptides fused to the amino terminus of the phage coat protein pIII (16,27). This results in a peptide fused to each of the three to five copies of pIII present on the phage particle. Short peptides (<10 amino acids) can also be displayed at the N-terminus of the major coat protein pVIII (28). The larger number of peptides displayed with the pVIII systems, approximately 2700 copies per phage particle, allows for the isolation of lower affinity interactions between the peptide and the target.

The utility of pIII and pVIII systems has been extended by the development of C-terminal display systems for each phage coat protein. Fuh *et al.* described the addition of a linker to the carboxyl terminus of pVIII thereby allowing the display of peptides fused to the end of the pVIII protein (29). This C-terminal pVIII phagemid system was successfully used to identify peptides that interact with PDZ domains. Modification of the pIII protein has also allowed the display of polypeptides fused to the carboxy-terminus of pIII (30).

Incorporation of the peptide-encoding region into the phage genome, as done in the pIII and pVIII display systems, is simple, works well, and is very useful for the identification of peptide ligands. These systems, however, are not very useful for larger polypeptides where the fusions often interfere with the growth or infectivity of the phage. The detrimental effects on growth of displaying a polypeptide on every pIII or pVIII can be overcome by the use of two-gene phagemid systems, where the polypeptide is fused to a phage coat protein in the phagemid vector and wild-type coat protein is supplied by a helper phage.

While pIII or pVIII display systems continue to be most popular, other filamentous phage coat proteins have been shown to be useful for display and selection. Jespers *et al.* demonstrated that a cDNA expression library can be fused to the C-terminus of the pVI protein (31). The pVI display system has also been used to identify immunogenic polypeptide ligands by cloning cDNA repertoires from tumor cells and then performing selections with antibodies (32). Phage coat proteins pVII and pIX have been used to display antibody fragments (33).

The utility of the filamentous phage display systems has led to the development of display systems with other bacteriophage. Bacteriophage lambda, display systems have been developed using the V protein, a major protein in the bacteriophage tail, and using the D protein, which is the major head protein (34–37). A phage display system based on bacteriophage T7 has also been developed where peptides or proteins are fused to the major capsid protein, gene 10 protein (38). The lytic phage, both lambda and T7, are well suited for the display of proteins or peptides that require cytoplasmic expression and cannot be secreted through the membrane as occurs in the filamentous phage display systems.

In addition to displaying peptides on the surface of bacteriophage particles, cell-free systems where the peptide remains linked to the mRNA encoding the peptide have been developed. Ribosome display involves *in vitro* translation of the mRNA under conditions that maintain the integrity of the nascent polypeptide–mRNA–ribosome complex. Affinity selections for peptide binding to the target are carried out using the complex, the mRNA is then eluted from the target, converted to cDNA, amplified, and used to initiate a new cycle of transcription/translation (39,40). A modification of the ribosome display method in which a covalent linkage is made between the mRNA and the polypeptide using puromycin was reported by Roberts and Szostack (41). Using this mRNA display system, Wilson and coworkers (42) were able to identify peptides with nanomolar affinity for streptavidin. The primary advantage of the cell-free systems over cell-based systems is the ability to produce libraries that are several orders of magnitude larger in size.

PEPTIDES AS TOOLS FOR DRUG DISCOVERY

Intracellular Expression of Peptides for Validation

The binding of peptides identified using display technologies can be validated prior to use in HTS as being critical for the function of the target. This is especially important when taking advantage of targets of unknown or putative function identified through genomic technology. To apply this approach, a peptide ligand to the protein target is isolated. The peptide is then expressed inside the cell, and the biological consequences are monitored. This approach was demonstrated in bacterial cells using peptides that were affinity selected by Novalon from peptide libraries displayed on phage for the prolyl-tRNA synthetase (12). Scientists at Cubist then expressed these peptides inside bacterial cells and demonstrated that the peptides inhibit growth both *in vitro* and in an *in vivo* mouse infection model (43). This method has since been demonstrated on multiple essential bacterial targets of known function and has been extended to allow for the selection of peptides by intracellular expression (15).

Figure 1 shows the results when peptides selected using phage display for three essential *E. coli* targets were expressed as fusions to glutathione-*S*-transferase (GST) under the control of a tightly regulated promoter. Induced expression of the peptide-GST fusion inhibited the growth of the bacterial cells while expression of GST alone did not. The target-specific nature of the growth inhibition

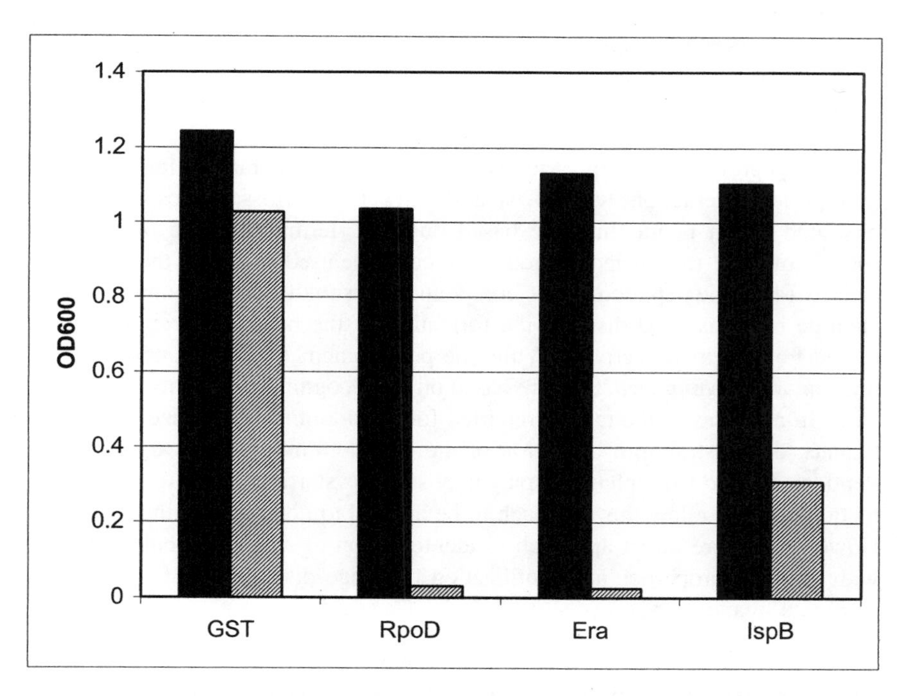

Figure 1 Inhibition of bacterial growth by intracellular expression of peptides selected by phage display that target essential bacterial proteins. Solid bars represent the optical density at 600 nm of an uninduced culture and hatched bars represent the optical density of the induced cultures. Glutathione-*S*-transferase is the fusion carrier protein and shows no significant reduction in growth while peptides isolated to RpoD, Era, and IspB show reduced growth upon induction of peptide expression.

could be demonstrated by coexpression of supplemental target protein in the cell, which resulted in relief of the inhibitory effect of the expressed peptide (15).

Peptides identified by display technology can also be used as tools for target validation. If a target is essential for growth of the bacteria, a peptide that binds to a functional site will inhibit the function of the target and inhibit bacterial growth. This provides target validation at the protein level rather than at the DNA or RNA level. Validation at the DNA level can be hampered by polar effects producing erroneous results. Since most drugs interact with proteins, validation through inhibition of a protein is a more direct demonstration of the validity of a target for drug discovery.

Peptidomimetic Chemistry. One approach that is often considered for the use of peptides in drug discovery is peptidomimetic chemistry. Peptides make unsuitable drugs in many instances due to poor oral availability and instability in serum (44). Peptidomimetic approaches begin with a peptide ligand that binds to the drug target and proceed through modification of the peptide to increase the "druglike" properties of the peptide. Development of inhibitors for proteases (45,46) and protein farnesyltransferase (47) have been reported using a peptidomimetic approach.

A peptidomimetic approach was also used to generate inhibitors of protein–protein interactions with *in vivo* antiviral activity. Moss and coworkers demonstrated that a peptidomimetic based on the C-terminal region of the small subunit of HSV ribonucleotide reductase could be used to inhibit the enzymatic activity (48). The peptidomimetic compound binds to the large subunit of ribonucleotide reductase and disrupts the formation of the heterodimeric enzyme required for enzymatic activity. While the peptidomemetic compounds discussed thus far are active *in vivo*, they are based on the recognition elements of peptides.

In most cases, the targets reported for peptidomimetics utilize peptides as ligands, demonstrating a limitation of the peptidomimetic chemistry approach. Peptides derived from phage display may serve as starting points for peptidomimetic design to allow this approach to be applied to nonpeptide binding targets. However, a more direct approach to identification of small-molecule inhibitors with druglike properties is the utilization of phage display-derived peptides as surrogate ligands for HTS.

Surrogate Ligand-Based Assays. Peptides have been extensively used for ligand binding assays and many methods have been developed to detect formation of the peptide–target complex. G-protein coupled receptors that utilize peptides as agonists are routinely screened using ligand binding assays with radioactivity, luminescence, or fluorescence for detection (49–53). These assays are simple and can be adapted to run on most HTS hardware. Therefore, the use of peptides from phage display as surrogate ligands is an extension of this concept that allows many target types to be assayed using established screening and detection methods.

Phage that display target-specific peptides can be used directly in an ELISA format to detect small-molecule inhibitors of a target. Phages isolated with *E. coli* dihydrofolate reductase (DHFR) were used to detect trimethoprim (TRM) or methotrexate (MTX), which are known inhibitors of DHFR (54). Figure 2 shows the results of incubating TRM or MTX with DHFR and a control protein and detecting the binding of the phage using and anti-M13 antibody-HRP conjugate. The binding of the control phage to its cognate target protein was not affected by TRM or MTX, while the binding of the DHFR phage to DHFR was dramatically reduced. While this method successfully detects small-molecule inhibitors, the use of M-13 phage as a reagent presents problems for HTS due to multivalent presentation of the binding peptides and day-to-day variations during amplification of the phage. The reproducibility required for HTS is provided by utilization of synthetic peptides rather than phage.

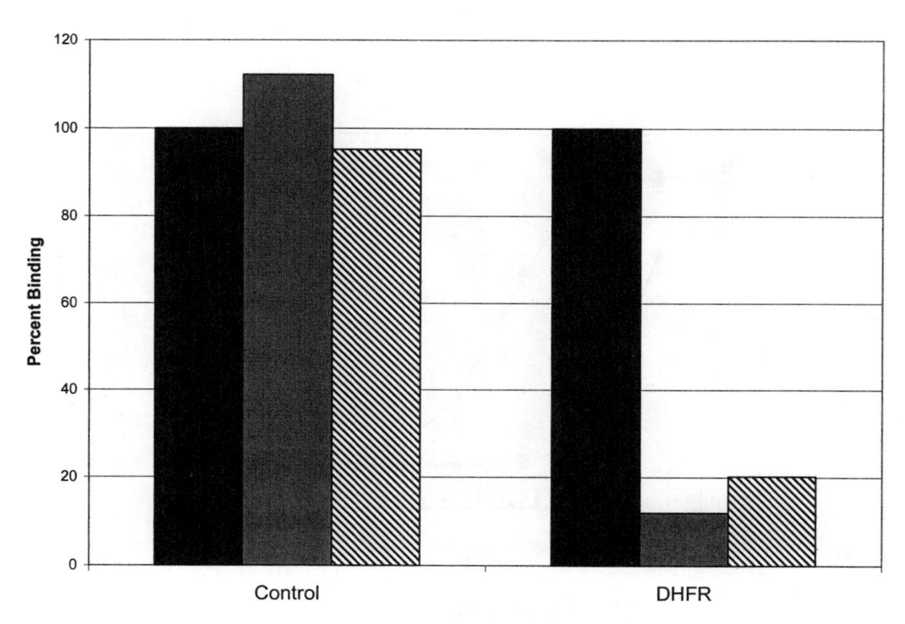

FIGURE 2 Phage binding to DHFR or a control protein in the absence of compound (black bars), methotrexate (gray bars), or trimethoprim (hatched bars). No reduction in phage binding occurs for the control protein while a significant reduction in binding occurs with the DHFR specific phage.

The use of synthetic peptides to detect inhibitors has been tested with both homogeneous and heterogeneous formats and detection, including radioactivity, luminescence, time-resolved fluorescence (TRF), fluorescence polarization (FP), and fluorescence resonance energy transfer. The titration curves for TRM and MTX, shown in Figure 3, were determined using a heterogeneous Peptide-on-Plate TRF assay (POP-TRF), where the biotinylated synthetic peptide is immobilized on a streptavidin coated microtiter plate, the remaining biotin binding sites were blocked with free biotin, and the plate was washed to remove unbound peptide and excess biotin. A conjugate was prepared with europium-labeled streptavidin and chemically biotinylated DHFR, and biotin was added to block all remaining biotin binding sites. The europium-labeled DHFR was incubated with the compounds and was transferred to the peptide-coated plate. After incubation to allow for binding, the microtiter plate was washed and the captured europium signal was detected.

To extend this work, four small-molecule inhibitors of tyrosyl-tRNA synthetase and a control compound were used to analyze several formats of the peptide competition assay (12). The titration curves of these compounds determined using the POP-TRF assay are shown in Figure 4. The titration curves show that compounds can be detected with a wide range of potencies.

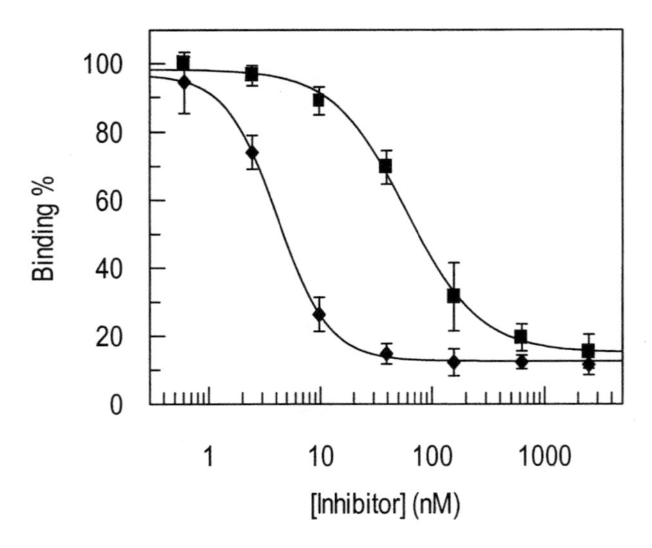

FIGURE 3 Titration curves for methotrexate (diamonds) and trimethoprim (squares) in peptide competition assays.

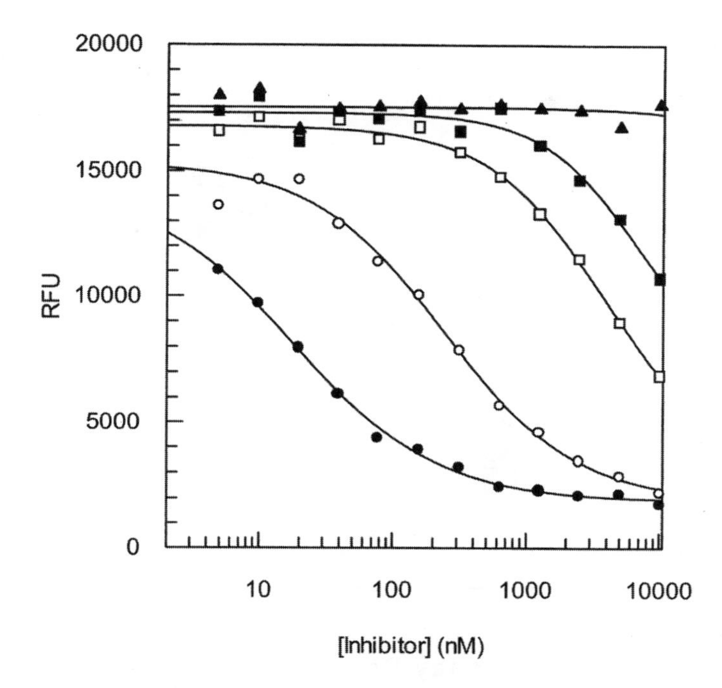

Figure 4 Titration curves of four small molecule inhibitors [NPC-0101 (solid circles), NPC-0102 (open circles), NPC-0103 (open squares), NPC-0104 (solid squares)], of tyrosyl-tRNA synthetase, and a control compound (solid triangles) as determined in a peptide competition assay.

Additional assay methods were tested to determine if the inhibitors would inhibit with the same potency in other assay formats and if the potencies would match the potencies observed in a biochemical assay. The concentration of inhibitor required to reduce the signal by 50% (IC_{50}) was determined for each inhibitor using a tRNA charging assay and five peptide-based surrogate ligand assay formats. A correlation plot of the observed IC_{50} values in the peptide assay versus the IC_{50} of the biochemical assay for each compound is shown in Figure 5. The potency of the compounds in each assay format remained consistent between assay formats, demonstrating the versatility of assay formats using phage-displayed peptides as surrogate ligands for HTS. Moreover, the correlation between potency in the biochemical assay and potency in peptide-based assays validates the use of these assays during compound optimization for targets of unknown function.

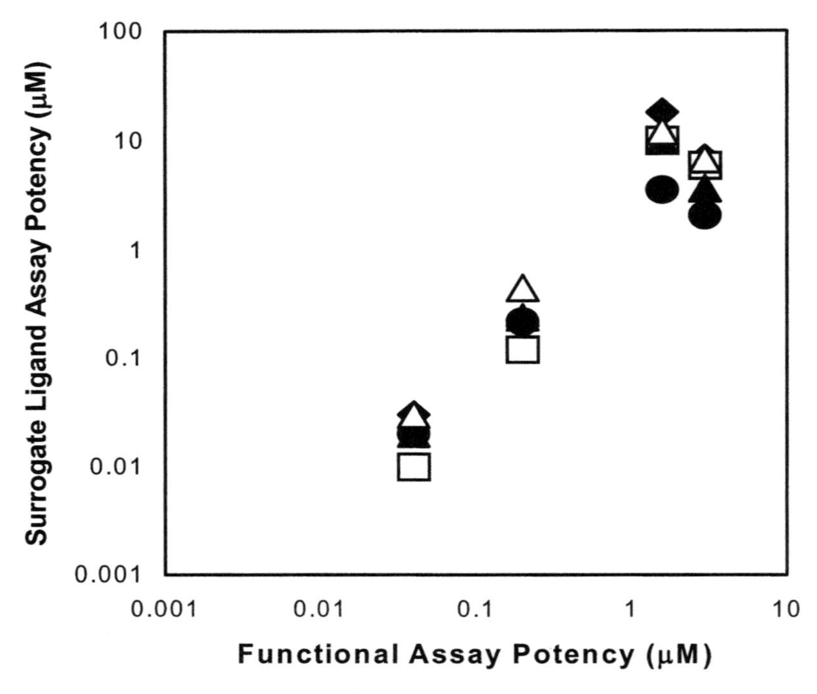

Figure 5 Correlation of the potencies of tyrosyl-tRNA synthetase from surrogate ligand-based assays and the biochemical assay.

CONCLUSION

Development of novel antibacterial compounds must be pursued continually to combat drug resistance. A large number of new targets for antibacterial drug discovery has been derived from whole-genome sequencing of bacterial species, but a large percentage of the genes encode proteins with no known biological function. One method for discovery of inhibitors of these targets of unknown function is surrogate ligand-based assays using molecular probes isolated through the use of display technologies. This provides a powerful method to screen and validate these targets. Advantages of surrogate ligand-based assay systems include flexibility when choosing an assay format and detection technology and the ability to validate a screening assay prior to the investment in full scale HTS. The use of this technology for the discovery of new antibacterial leads has been reported (55–58) and provides a novel approach that is especially useful for drug discovery efforts aimed at targets of unknown function.

REFERENCES

1. DTW Chu, *et al.* New directions in antibacterial research. *J Med Chem 39*: 3853–3874, 1996.
2. RD Fleischmann *et al.* Whole-genome random sequencing and assembly of *Haemophilus influenzae* Rd. *Science* 269:496–512.
3. FR Blattner *et al.* The complete genome sequence of *Escherichia coli* K-12, *Science 277*:1453–1474, 1997.
4. F Kunst *et al.* The complete genome sequence of the gram-positive bacterium *Bacillus subtilis. Nature 390*:249–256, 1997.
5. H Tettelin *et al.* Complete genome sequence of *Neisseria meningitidis* serogroup B strain MC58. *Science 287*:1809–1815, 2000.
6. CK Stover *et al.* Complete genome sequence of *Pseudomonas aeruginosa* PA01, an opportunistic pathogen. *Nature 406*:959–964, 2000.
7. J Joseph *et al.* Complete genome sequence of an M1 strain of *Streptococcus pyogenes. Proc Natl Acad Sci USA 98*:4658–4663, 2001.
8. J Trias, EM Gordon. Innovative approaches to novel antibacterial drug discovery. *Curr Opin Biotechnol 8*:757–762, 1997.
9. MW Pantoliano *et al.* Microplate thermal shift assay apparatus for ligand development and multi-variable protein chemistry optimization, US 6,036,920, 2000.
10. JU Bowie, AA Pakula. Scriptgen Pharmaceuticals screening method for identifying ligands for target proteins, US 5,585,277, 1996.
11. DE Hughes *et al.* Capillary electrophoretic method to detect target-binding ligands and to determine their relative affinities, WO 99/34203, 1999.
12. R Hyde-DeRuyscher, *et al.* Detection of small-molecule enzyme inhibitors with peptides isolated from phage-displayed combinatorial peptide libraries. *Chem Biol 7*:17–25, 2000.
13. DJ Christensen *et al.* Phage display for target-based antibacterial drug discovery. *Drug Discov Today 6*:721–727, 2001.
14. J Tao *et al.* Drug target validation: lethal infection blocked by inducible peptide. *Proc Natl Acad Sci USA 97*:783–786, 2000.
15. RE Benson *et al.* Intracellular validation of surrogate ligands for antimicrobial drug discovery. 101st General Meeting of the American Society of Microbiology, Orlando, FL, 2001.
16. GP Smith. Filamentous fusion phage: novel expression vectors that display cloned antigens on the virion surface. *Science 228*:1315–1317, 1985.
17. GP Smith, VA Petrenko. Phage display. *Chem Rev 97*:391–410, 1997.
18. SE Cwirla *et al.* Peptides on phage: a vast library of peptides for identifying ligands. *Proc Natl Acad Sci USA 87*:6378–6382, 1990.
19. JJ Devlin *et al.* Random peptide libraries: a source of specific protein binding molecules. *Science 249*:404–406, 1990.
20. JK Scott, GP Smith. Searching for peptide ligands with an epitope library. *Science 249*:386–390, 1990.
21. H Grøn, RP Hyde-DeRuyscher. Peptides as tools in drug discovery. *Curr Opin Drug Discov 3*:636–645, 2000.

22. GK Ehrlich, P Bailon. Identification of peptides that bind to the constant region of a humanized IgGI monoclonal antibody using phage display. *J Mol Recognit 11*: 121–125, 1998.
23. GK Ehrlich *et al.* Phage display technology: Identification of peptides as model ligands for affinity chromatography. *Methods Mol Biol 147*:209–220, 2000.
24. BK Kay *et al.* Convergent evolution with combinatorial peptides. *FEBS Lett. 480*: 55–62, 2000.
25. R Pasqualini, E Ruoslahti. Organ targeting *in vivo* using phage display peptide libraries. *Nature 380*:364–366, 1996.
26. R Pasqualini, *et al.* Aminopeptidase N is a receptor for tumor-homing peptides and a target for inhibiting angiogenesis. *Cancer Res 60*:722–727, 2000.
27. SF Parmley, GP Smith. Filamentous fusion phage cloning vectors for the study of epitopes and design of vaccines. *Adv Exp Med Biol 251*:215–218, 1989.
28. J Greenwood *et al.* Multiple display of foreign peptides on a filamentous bacteriophage: peptides from *Plasmodium falciparum* circumsporozoite protein as antigens. *J Mol Biol 220*:821–827, 1991.
29. G Fuh *et al.* Analysis of PDZ domain-ligand interactions using carboxyl-terminal phage display. *J Biol Chem 275*:21486–21491, 2000.
30. G Fuh, SS Sidhu. Efficient phage display of polypeptides fused to the carboxy-terminus of the M13 gene-3 minor coat protein. *FEBS Lett 480*:231–234, 2000.
31. LS Jespers *et al.* Surface expression and ligand-based selection of cDNAs fused to filamentous phage gene VI. *Biotechnology 13*:378–382, 1995.
32. SE Hufton *et al.* Phage display of cDNA repertoires: the pVI display system and its applications for the selection of immunogenic ligands. *J Immunol Method 231*:39–51, 1999.
33. C Gao *et al.* Making artificial antibodies: a format for phage display of combinatorial heterodimeric arrays. *Proc Natl Acad Sci USA 96*:6025–6030, 1999.
34. IN Maruyama *et al.* Lambda foo: a lambda phage vector for the expression of foreign proteins. *Proc Natl Acad Sci USA 91*:8273–8277, 1994.
35. N Sternberg, RH Hoess. Display of peptides and proteins on the surface of bacteriophage lambda. *Proc Natl Acad Sci USA 92*:1609–1613, 1995.
36. IS Dunn. Assembly of functional bacteriophage lambda virions incorporating C-terminal peptide or protein fusions with the major tail protein. *J Mol Biol 248*: 497–506, 1995.
37. YG Mikawa *et al.* Surface display of proteins on bacteriophage lambda heads. *J Mol Biol 262*:21–30, 1996.
38. A Rosenberg *et al.* T7Select phage display system: a powerful new protein display system based on bacteriophage T7. *Innovations (Newslett Novagen) 6*:1–6.
39. LC Mattheakis *et al.* An in vitro polysome display system for identifying ligands from very large peptide libraries. *Proc Natl Acad Sci USA 91*:9022–9026, 1994.
40. GM Gersuk *et al.* High-affinity peptide ligands to prostate-specific antigen identified by polysome selection. *Biochem Biophys Res Commun 232*:578–582, 1997.
41. RW Roberts, JW Szostak. RNA-peptide fusions for the in vitro selection of peptides and proteins. *Proc Natl Acad Sci USA 94*:12297–12302, 1997.
42. DS Wilson *et al.* The use of mRNA display to select high-affinity protein-binding peptides. *Proc Natl Acad Sci USA 98*:3750–3755, 2001.

43. JH Blum *et al.* Isolation of peptide aptamers that inhibit intracellular processes, *Proc Natl Acad Sci USA 97*:2241–2246, 2000.
44. MD Taylor, GL Amidon, eds. *Peptide-Based Drug Design*. Washington, DC: American Chemical Society, 1995.
45. SK Thompson *et al.* Structure-based design of cathepsin K inhibitors containing a benzoyl peptidomimetic. *J Med Chem 41*:3923–3927, 1998.
46. PS Dragovich *et al.* Structure-based design, synthesis, and biological evaluation of irreversible human rhinovirus 3C protease inhibitors. 3. Structure–activity studies of ketomethylene-containing peptidomimetics. *J Med Chem 42*:1203–1212, 1999.
47. SJ O'Connor *et al.* Second generation peptidomimetic inhibitors of protein farnesyltransferase demonstrating improved cellular potency and significant *in vivo* efficacy. *J Med Chem 42*:3701–3710.
48. N Moss *et al.* Peptidomimetic inhibitors of herpes simplex virus ribonucleotide reductase with improved in vivo antiviral activity. *J Med Chem 39*:4173–4180, 1996.
49. BL Daugherty *et al.* Radiolabeled chemokine binding assays. *Methods Mol Biol 138*: 129–134, 2000.
50. GJ Parker *et al.* Development of high thorughput screening assays using fluorescence polarization: nuclear receptor-ligand-binding and kinase/phosphatase assays. *J Biomol Screen 5*:77–88, 2000.
51. KC Appell *et al.* Biological characterization of neurokinin antagonists discovered through screening of a combinatorial library. *J Biomol Screen 3*:19–27, 1998.
52. KJ Valenzano *et al.* Development of a fluorescent ligand-binding assay using the acrowell filter plate. *J Biomol Screen 5*:455–461, 2000.
53. M Allen *et al.* High throughput fluorescence polarization: A homogeneous alternative to radioligand binding for cell surface receptors. *J Biomol Screen 5*:63–69, 2000.
54. B Kay, PT Hamilton. Identification of enzyme inhibitors from phage-displayed combinatorial peptide libraries. *Comb Chem High Throughput Screen 4*:535–543, 2001.
55. J Finn *et al.* Enhancing drug discovery: utilization of VITA fluorescently labeled ligands in high throughput capillary electrophoresis screening. 40th Interscience Conference on Antimicrobial Agents and Chemotherapy, Toronto, On, Canada (Abstract 2031), 2000.
56. EB Gottlin *et al.* Surrogate ligands in antibacterial drug discovery. 40th Interscience Conference on Antimicrobial Agents and Chemotherapy, Toronto, ON, Canada (Abstract 2032), 2000.
57. DJ Christensen *et al.* Surrogate ligand-baseds high throughput screening of difficult-to-assay targets. 41st Interscience Conference on Antimicrobial Agents and Chemotherapy, Chicago, IL (Abstract 2131), 2001.
58. A Jacobi *et al.* Isoprenoid biosynthesis as a novel antibacterial target. 41st Interscience Conference on Antimicrobial Agents and Chemotherapy, Chicago, IL (Abstract 2124), 2001.

12

Expression Profiling Uses in Antibacterial Chemotherapy Development

Paul M. Dunman and Steven J. Projan
Wyeth Research, Pearl River, New York, U.S.A.

INTRODUCTION

Conventionally, antibacterial drug development projects have been geared toward inhibiting bacterial biosynthetic mechanisms that are (1) essential for prokaryotic survival, (2) unique to the pathogen as opposed to humans, and (3) conserved across bacterial species (to ensure broadspectrum activity). Many classes of antibiotics have been developed based on these criteria, including β-lactams and aminoglycosides. Collectively these antibiotics have offered a significant level of effectiveness against pathogens. However, as levels of antibacterial resistance continue to increase among bacterial species (for example, clinical isolates of *Staphylococcus aureus* have demonstrated resistance to all known currently available antibiotics), it is apparent that novel microbial enzymatic processes need to be targeted for drug development programs.

The advent of comparative genomics has presented new opportunities for identifying potential antimicrobial targets. At the time of writing this text, perusal of the Institute for Genomic Research Web site (TIGR; **www.tigr.org**) provides complete genomic sequence information for 35 bacterial species and partially completed data for another 126 organisms. Undoubtedly these numbers will increase dramatically by the time of publication, a testament not to the laborious publication procedure, but rather to the speed at which bacterial genomes are being delineated.

As this information becomes available researchers can scan for genes that are conserved across prokaryotes yet have limited or no significant homology to sequences of the recently published human genome. Candidate target genes can then be assayed for essentiality using standard molecular techniques. However, determining whether a particular gene is essential, and thus an attractive antibacterial drug target, is not a trivial task. Procedures to determine essentiality must be well thought out, or potential targets may be missed, as in the case of gene products that are not required for *in vitro* bacterial growth but are required for *in vivo* survival (and *vice versa*). Likewise, results of essentiality tests may be initially misleading. For example, the *Escherichla coli uvrD* and *rep* genes encode for helicases that are functionally redundant. A knockout in either gene produces little effect on bacterial growth; however, the double mutant is not viable. In *Staphylococcus aureus*, the *uvrD/rep* homolog is the *pcrA* gene, a gene that is essential for growth *in vitro* (i.e., there is no redundancy in *S. aureus* for this essential helicase). The relative similarity between *rep* and *uvrD* (and *pcrA*) suggests that an antibacterial compound could be developed which successfully targets all three enzymes. As a result, although neither *rep* nor *uvrD* is "essential" in *E. coli* they constitute a functionally essential enzymatic activity and are valid targets for classical antibacterial drug discovery (1). Conversely there have been several reports describing "essential" two component regulatory systems (TCRS). Yet the precise essential function of these TCRS have yet to be elucidated and may merely reflect the toxic derepression of genes negatively, as opposed to genes positively regulated by the system (2–4). Clearly, defining essentiality (and thus a good candidate for antibacterial drug development) is a labor-intensive task that requires an intimate understanding of microbial physiology.

Identification of gene products that are novel, unique, conserved, and "essential" among bacteria is only the beginning steps of identifying a putative antibacterial target. Most drug development programs are function-based, and therefore characterization of the gene product is a prerequisite for it to be considered a valid target for anti-infective development. Despite the outpouring of bacterial genomic data available, understanding the function(s) of genes constituting these organisms has lagged behind and constitutes the major bottleneck in novel antibacterial drug development projects. It has been estimated that more than one-third of the genes of the best studied organisms, such as *E. coli*, have no defined function (5). Accordingly, researchers are turning to high-throughout methodologies to help assign function to modestly characterized gene products.

Expression profiling technologies allow one to gain information about the expression behavior of uncharacterized genes in prioritizing target validation efforts. Combined with traditional studies, these procedures also promise to help bridge the gap between gene sequence evaluation and gene product function. The underlying assumption is that, because most bacterial genes are expressed on an as-needed basis, identifying the levels at which and when gene products are produced allow correlations to be made between the functions of genes sharing similar expression patterns. To this end, by comparing more than 2000 *Saccharomyces cerevisiae* gene expression patterns across 78 experiments investigating 8 biological processes, Brown, Botstein, and coworkers have shown that genes sharing expression patterns often share function (6). Once a hypothesis has been established about the function of a gene product, it can subsequently be tested by creating mutations in the gene and studying the corresponding phenotype(s). We first describe some of the current methodologies for bacterial expression profiling and then discuss the utility of such technologies in antibacterial target identification, assay development, and validation.

TRANSCRIPTION PROFILING

Transcription profiling provides an unprecedented ability to monitor genomewide differences in gene transcription. Several transcription profiling methods have been developed; however, two procedures are most commonly used, DNA microarrays and oligonucleotide arrays. In general these techniques involve labeling an RNA population, which is subsequently hybridized to a solid matrix containing representatives of open reading frames in defined locations (quadrants). The matrix is then assayed for RNA hybridized to each quadrant. The amount of RNA detected indicates the steady-state level of that transcript in the RNA sample being tested. Such arrays have seen extensive use in eukaryotic biology and are only recently being used to study prokaryotic processes. The lack of utility in bacterial samples is due, in part, to the absence of robust prokaryotic RNA labeling technologies. In contrast, eukaryotic RNA samples are commonly labeled by synthesizing fluorescent probes for mRNA using reverse transcriptase and oligo(dT) primers containing RNA polymerase promoters, followed by *in vitro* RNA polymerase reactions. Similar procedures have recently been adapted to label bacterial samples. Investigators are also using techniques to directly label bacterial RNA material, albeit with much lower specific activity. Moreover, most bacterial mRNAs are thought to be much less stable than eukaryotic messages *in vivo*, complicating the isolation of usable, high-integrity RNA.

DNA Microarrays

Although there are several derivations, DNA microarrays are usually constructed by affixing members of a genomic library or PCR products to a solid surface; such as a glass slide, *via* a robotic spotting device [reviewed in (7)]. Each member is attached (tiled) within a given quadrant (element) of the microarray. Typically, RNA prepared from two samples that are being compared are differentially la-

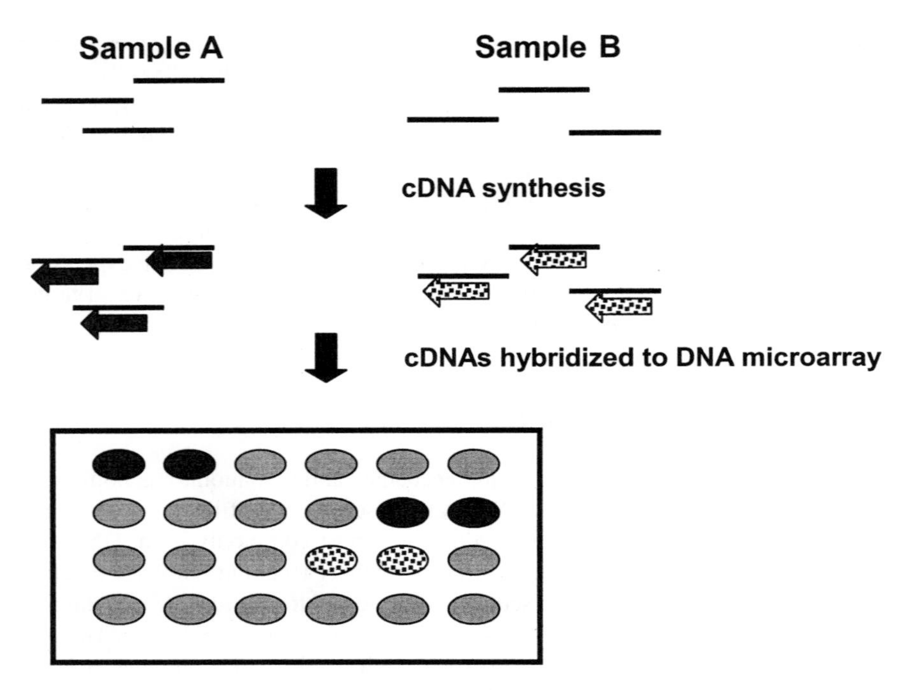

Figure 1 DNA microarray overview. RNA samples are differentially labeled and hybridized onto a matrix that typically harbors PCR products of predicted open reading frames.

beled with fluorescent moieties (Figure 1). One RNA sample serves as a template for cDNA synthesis, using either random hexamers or oligo(dT) as primers, during which Cy5- (a red fluorochrome) labeled nucleotides are incorporated into the newly synthesized strand by reverse transcriptase. The second RNA sample is labeled in a similar manner, using Cy3- (green fluorochrome) labeled nucleotides. The two labeled cDNA samples (probes) are then mixed and hybridized onto the microarray. The array is scanned to detect Cy5- and Cy3-fluorescing elements on the microarray. Cy5 and Cy3 signals are superimposed; yellow signals indicate an equal amount of RNA from each sample has hybridized to an attached library or PCR species. A red or green signal indicates a predominance of transcript in the Cy5- or Cy3-labeled RNA sample, respectively.

Recently, investigators have used DNA microarrays to define genes involved in bacterial processes such as *Bacillus subtilis* sporulation (8), *Escherichia coli* nitrogen limitation (9), and *Streptococcus pneumoniae* competence (10). In those studies results confirmed expression patterns of genes known to be involved in the biological process examined and assigned function to previously uncharacterized genes products. More recently, several investigators have used this technology to define the regulation of host genes in response to bacterial infection [reviewed by (11)]. Collectively these studies have provided insight into how mammalian cells respond to infection, producing information that may ultimately be parlayed into antibacterial therapeutics. Premade DNA microarrays are commercially available from several companies, including Clontech (Palo Alto, CA) and Incyte Pharmaceuticals (Palo Alto, CA). Alternatively, "homemade" arrays can be made using technology pioneered by Pat Brown and colleagues (Stanford University, Stanford, CA). A number of factors have contributed to the popularity of using DNA microarrays to study global transcription patterns, including (1) they tend to be inexpensive relative to other technologies (2) they do not necessarily require the DNA sequence of samples tiled onto the microarray to be previously determined, and (3) they allow the investigator a great deal of flexibility in selecting the clones they choose to attach onto the DNA array. However, DNA arrays are restricted by their sensitivity limitations in comparison to other technologies and they are prone to ambiguous results due to cross-hybridization of cDNA species, such as small fragments (7). Additionally, due to the polycistronic nature of bacterial mRNA, a single probe may partially extend into adjacent ORFs. Moreover, probes (of various sizes) may have differences in hybridization temperatures, complicating analyses.

Oligonucleotide Arrays

Oligonucleotide arrays are commercially available from companies such as Affymetrix [reviewed in (12)]. Affymetrix uses photolithography technology to synthesize several (typically 20–25) oligonucleotides (perfect match) that are complementary to mRNA of predicted ORFs of an organism under investigation on the surface of a GeneChip. Additionally, mismatch oligonucleotides that are constructed with nearly identical sequence to perfect match oligonucleotides, differing only in a central residue, are also synthesized (tiled) onto the array. Total bacterial RNA is extracted from a bacterial population (Figure 2). It is enriched for mRNA by the addition of rRNA specific oligonucleotides that serve as primers for a cDNA synthesis step, which is followed by sequentially adding RNAse H and DNAse I to remove rRNA/cDNA molecules. Following mRNA enrichment, transcripts are fragmented and 5′ biotinylated. The RNA is then hybridized to oligos on the GeneChip and stained by streptavidin conjugated to a fluorescent moiety, such as phycoerythrin. Scanning confocal microscopy allows for fluorescent intensity values to be determined for RNA species annealed to both perfect and mismatch oligonucleotides. Average difference values for

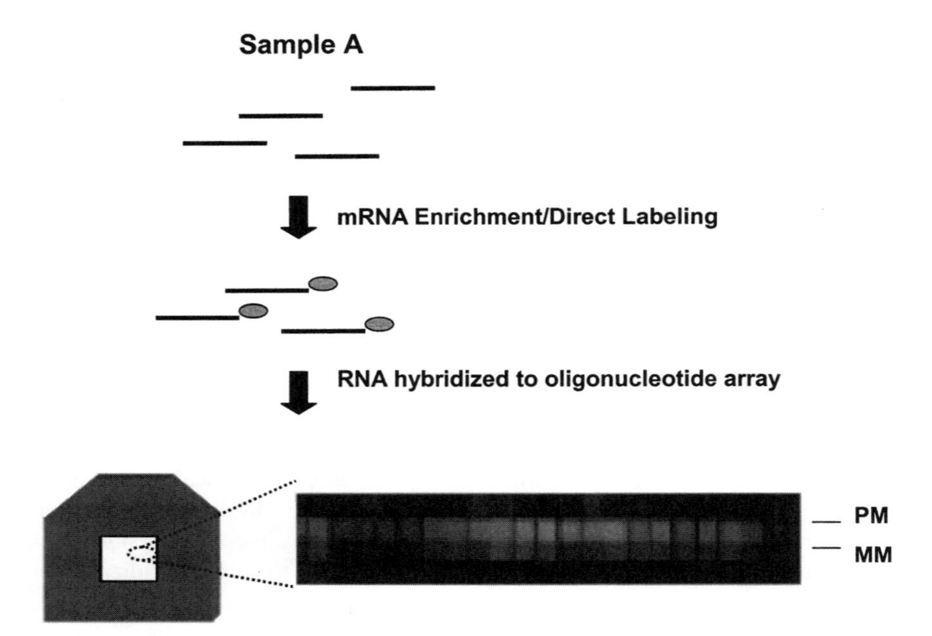

Figure 2 Oligonucleotide array overview. The RNA sample is labeled and hybridized onto a matrix harboring 25 perfect-match (PM) and 25 mismatch (MM) oligonucleotides (25-mers) derived from predicted open reading frames. Shown (bottom right) is RNA hybridized to PM and MM oligonucleotides of a gene, after scanning for flourescently labeled RNA.

ORFs are determined after subtraction of mismatch from perfect-match oligonucleotide signals. Additionally, Affymetrix algorithums calculate whether a given transcript is considered "present" within a RNA sample. This technology is generally regarded as being more sensitive than typical DNA microarrays, although direct comparative studies using bacterial RNA samples have not been adequately reported in the literature. The specificity of transcript signal, as determined by comparing mismatch and perfect match oligonucleotide signals, is likely to contribute to the sensitivity of the system.

Despite labeling issues investigators have successfully used oligonucleotide arrays to identify genes involved in bacterial processes such as *Staphylococcus aureus* virulence (13), *S. aureus* oxacillin resistance (14), and *Streptococcus pneumoniae* quorum-sensing processes (15). Drawbacks to using oligonucleotide arrays include that they are generally more expensive than DNA microarrays and that the sequence of the organism to which the array is being constructed must be previously determined. Additionally, in comparison to cDNA-labeling procedures, direct labeling approaches require large amounts of starting material, which may be problematic in some experimental systems. Moreover, incorporation of multiple fluorochromes (as in the cDNA labeling procedures) yields much higher signal strength per probe than a direct end-labeling method. However, improved labeling procedures are being developed that will circumvent many of these shortcomings.

PROTEOMICS

Although DNA and oligonucleotide arrays have gained popularity among researchers, one cannot discount that transcript levels do not necessarily directly correlate to protein abundance within the cell or measure the effects of post translational modifications to protein components of the cell. As in the case of studying transcript titers, establishing a protein's abundance within a sample is expected to further allow inference to the protein's function(s) in biological processes, which is the basis of proteomics (reviewed in 16–18).

Two-Dimensional Polyacrylamide Gel Electrophoresis

Two-dimensional polyacrylamide gel electrophoresis (2D-PAGE) allows for separating labeled proteins of a sample in one dimension on the basis of charge and then in a second dimension by size (18). The resulting gel can be analyzed for the location of labeled spots. Comparison of the patterns of labeled proteins for bacteria subjected to different variables allows for the determination of spots, or proteins, that are due to the experimental variable. Proteins of interest can be subsequently isolated from the gel and directly analyzed by mass spectrometry or by chemical (or enzymatic) degradation and protein sequencing. Resulting amino acid sequence information can then be used to mine available databases for identification of the protein. This approach has been successfully used by researchers to identify proteins that are expressed in response to changes in environmental conditions, such as exposure to antibiotics (19). However, 2D-PAGE analysis combined with mass spectrometry has been shown to detect only the most abundant proteins (20). Additionally, although the number of spots typically observed in 2D-PAGE correlate well with predictions regarding protein species expected to be within a sample, the technology is inherently biased by the solubility, dynamic range, and focusing of proteins (18). Additionally, resolution of protein samples is not always complete; multiple proteins may comigrate, making identification of proteins within a spot difficult and reproducibility an issue (16). In one study Gauss et al. (21) that one gene can give rise to more than 500 discreet spots. Admittedly, that study did not involve bacterial samples; however, it is not likely that 2D-PAGE of prokaryotic proteins will be immune from such artifacts.

Proteomics on a Chip

Another recent approach has been to use chip-based technology to identify proteins within a sample (16). In this approach chips are prepared with different charges or antibodies or with nucleic acids on their surfaces. Protein samples can then be applied and separated based on their affinity for the surface. They are then identified by mass spectrophotometry. This technology is reminiscent of traditional protein chromatography and, at the current stage of use, should be considered as such. This methodology is likely to be very sensitive and applicable to certain studies, yet it suffers from the limitation that it will not provide a global analysis of a sample and will be restricted to the differentiation capabilities of the chip, *i.e.*, a cell surface protein may not readily bind to a chip with nucleic acids on its surface. Although other versions of proteomics on a chip are being developed, their utility is currently in the infant stages.

Collectively, the use of DNA (or oligonucleotide) microarrays and proteomics is likely to link functionally unknown proteins to biological processes, allowing a "guilt by association" characterization of gene products. In practice, this is likely to require very stringent experimental design and compilation of data from many studies, which will need to be confirmed by more classical approaches. Although researchers have used both techniques to identify genes that are likely to be involved in bacterial processes, such as *Bacillus subtilus* sporulation, *Staphylococcus aureus* antibiotic resistance, and *S. aureus* virulence, very few studies have yet to confirm these predictions.

TARGET IDENTIFICATION

Both the transcription profiling and proteomics fields provide information regarding expression patterns of genes; however, the usefulness of the data in ascribing function to previously uncharacterized genes remains to be seen. What is evident is that both techniques provide an overwhelming amount of data, which by itself is limited. However, when compared across a number of other experiments, profiling data is expected to result in a better correlation between cellular function and gene expression. For example, identifying genes regulated in response to agents that inhibit a process, such as cell division, will likely provide a list of genes which includes general stress responsive ORFs. Comparisons must then be made between these genes and those that are identified as being regulated by other stresses. The question then becomes how many additional stresses need to be evaluated? Additionally, small changes in the expression of a given gene may have more profound effects on the cell than more dramatic changes. How does one evaluate fold changes in gene expression? These are questions that the field is currently not yet ready to answer. Nevertheless, it is becoming more and more apparent that global expression patterns are a tool that investigators can use to help fill in gaps in the description of previously uncharacterized genes. Therefore, at the current level of use, profiling data alone is not able to solely assign function to genes.

DEVELOPMENT OF ANTIBACTERIAL SCREENS FROM PATHWAY AND TxP STUDIES

In addition to assisting in ascribing functions to unknown gene products, these technologies have also been employed in developing cell-based screens for antibacterial development. In this approach, the transcripts or proteins induced by inhibition of an essential bacterial process are identified. Promoters for these gene products are then attached to reporter genes, such as those encoding β-galactosidase and luciferase. Bacteria harboring these reporter systems are then exposed to sublethal concentrations of potential antibacterial agents. Subsequent expression of the reporter enzyme indicates that the biological process has been affected by the agent in question. Murphy and coworkers at Millennium Pharmaceuticals have taken this approach to identify genes upregulated in response to agents that inhibit processes such as fatty-acid biosynthesis. Likewise Dunman and colleagues (manuscript in preparation) have recently identified 27 genes that are induced by a variety of cell-wall active antibiotics, including oxacillin, bacitracin, vancomycin, and D-cycloserine (Table 1). Although powerful, there are several potential shortcomings to this type of approach. As indicated above, other types of stress may induce expression of the reporter gene used.

One obvious benefit of using a cell-based assay for antibacterial development is that the assay inherently identifies agents that are taken up by the cell. The downside is that agents that are potent inhibitors of a bacterial process but are not efficiently taken up the cell, but which could be chemically manipulated to do so, are not picked up in cell-based screens. Also, the window between detecting the reporter gene product and bacterial killing may be problematic.

MECHANISM OF ACTION STUDIES

Expression profiling is expected to aid in proving the mechanism for antibiotics developed to inhibit specific biological processes. By comparing transcript or protein profiles of cells that have been rendered deficient for a given process (such as through the controlled expression of the target gene) to agents that are suspected to have similar inhibitory activities, researchers can provide *in vivo* proof that an antibiotic is, indeed, inactivating that process within the cell. As shown in Table 1, cell-wall-active antibiotics induce expression of a network of genes within *S. aureus*. Compounds being developed for antibacterial chemotherapy which target cell-wall synthesis would therefore be expected to also induce these genes. However, caution should be applied when validating the mechanism

TABLE 1 Genes Induced by Cell-Wall-Active Antibiotics

		Fold induction[c]				
Chip ORF[a]	Genbank[b]	Bacitracin	Cycloserine	Vancomycin	Oxacillin	N315[d]
2372	*butA*	4.5 (<0.1)	6.1 (0.6)	3.9 (0.5)	2.2 (0.2)	SA0122
3990		2.2	2.5	2.9	2.7	SA0182
4587		5.3 (0.7)	5.5 (2.0)	2.7 (1.0)	2.7 (0.1)	SA0758
1863		2.6	5.7	3.9	2.3	SA0841
3773		2.3	3.3	2.6	2.6	SA0864
2058	*ribC*	2.5	3.6	2.3	2.1	SA1115
2515	*bsaA*	4.2 (2.8)	4.6 (3.2)	3.0 (1.0)	3.3 (0.1)	SA1146
4692	*katA*	3.6	2.5	2.1	2.2	SA1170
2642	*citB*	3.7	4.3	2.3	3.9	SA1184
1177		4.1	5.2	3.4	2.2	SA1192
2599		4.1	5.8	2.4	4.5	SA1215
4962	*crr*	9.6 (1.2)	9.3 (1.2)	2.3 (0.3)	7.8 (1.3)	SA1255
2167		7.8	8.1	2.4	7.6	SA1256
4964	*mrsA*	12.0	11.1	2.1	15.7	SA1257
1325		3.2	3.0	2.3	2.9	SA1295
5481		3.9	3.7	2.6	2.8	SA1490
5088		2.8	2.5	3.3	2.4	SA1543
5081	*htrA*	12.7	8.8	2.8	7.6	SA1549
4100		2.3	2.6	2.7	2.1	SA1606
4520	*dinP*	2.6 (0.5)	2.8 (0.5)	2.6 (0.2)	4.2 (<0.1)	SA1711
170		3.0	4.0	3.0	10.5	SA1988
3964		2.3	2.0	2.6	2.5	SA1989
2201		10.3	9.1	2.2	12.6	SA2103
4163		5.4	6.8	2.8	3.1	SA2139
2508		5.3	5.4	2.0	2.1	SA2297
5248	*copA*	6.2 (2.0)	3.7 (1.5)	3.1 (0.1)	9.2 (0.7)	SA2344
172		6.0	4.7	3.0	3.0	

[a] Designated *S. aureus* GeneChip open reading frame number.
[b] Previously described gene name.
[c] Fold induction when cells are treated with indicated antibiotic, as compared to mock treated cells. Standard deviation shown in parenthesis.
[d] Corresponding designated *S. aureus* strain N315 gene (*24*).

of action of a compound by these means. For example, cell-wall-active antibiotics that inhibit a different step of cell-wall synthesis (i.e., early vs. late stage of peptidoglycan synthesis) may yield an entirely different transcript profile.

Studies conducted with putative antibacterial agents administered to eukaryotic cells can be used to provide information regarding the toxicity of drugs in design. In this scenerio, it is expected that eukaryotic cells treated with sublethal doses of toxic agents will demonstrate distinct transcript and protein profiles. Extending this, potential antibiotics that demonstrate a similar profile may be toxic, eliminating time and money that would otherwise be spent on their development.

Correlating a transcription profile signature to a drug or genetic effect is highly complex, but has been used successfully in other fields. Butte *et al.* have used statistical analysis to correlate RNA expression to drug sensitivity among cell lines that respond to different anticancer drugs (22). This study examined 60 cell lines, 7245 genes, and 5024 drugs to describe the relationship between transcription profile and drug class. Using a refined set of 6701 genes and 4991 drugs, they described drug–profile relations through a system of relevance networks. An understanding of how antibiotics relate to expression profiles, particularly across bacterial species, will help in evaluating lead compounds, especially when the lead compound inhibits the target activity but does not display inhibitory activity or a compound that does show such activity, but may hit more than one target.

Protein arrays have also been developed for functional analysis. In this approach synthetic peptides or purified proteins are distributed onto a solid surface. Peptide arrays can subsequently be used in mapping interactions between proteins, a technique which is expected to be invaluable in developing antibiotics designed to inhibit similar interactions *in vivo* and can be viewed as the equivalent of performing an *in vitro* two-hybrid protein screen. Typically the peptides are synthesized directly on a solid support (as in the case with oligonucleotide arrays) using photoliabile protection groups on the peptide chain. A mask containing defined transparent locations is placed over the array and exposed to light, deprotecting the terminal amino acid of the peptide chain. The amino acid to be coupled to the growing chain is then added. Using various masks and the stepwise addition of specific amino acids discreet locations on the array surface will harbor synthesized polypeptide chains of known sequence. Although this technology routinely allows for synthesis of peptide chains on the order of 30 residues in length to be tiled onto a support surface, this is perceived as the upper limits of peptide length and a potential hurdle which must be overcome to study certain relevant protein–protein interactions. Other shortcomings of this technology are that unknown cofactors that may normally mediate protein–protein interactions may not be immediately available and may limit the effectiveness of the system. Additionally, the synthesized molecules may not form a conformation (fold) that would be expected *in vivo*.

PROFILING IN MODELS OF PATHOGENESIS

Expression profiling allows researchers to examine the process of invasion and pathogenesis without the bias of genetic selection. Although technically difficult, these studies promise new insights into the progression from microbial growth to infection. Israel *et al.* have recently examined host response to two different strains of *H. pylori* of differing virulence and then used a whole genome microarray to find that the determining factor for severity of virulence was the present of the *cag* gene (23). The use of expression profiling technologies can also detect host responses during and following infection and allow the identification of components of the defense systems which can be enhanced or mimicked to combat the infection (5). In some cases, human DNA microarrays may be utilized or, alternatively, microarrays for animals used in model studies could be advantageous. Because of the complexity of samples from actual tissues composed of different cell types, use of tissue culture may result in more specific, yet more artificial, analysis of a response. Beyond the use of these comparisons to yield potential drug targets, it has been speculated that unique transcription responses to different pathogens might aid in the development of diagnostic tools for clinical use to identify the infecting organism. While an attractive thought, given the common use of empirical treatment, this does not seem currently feasible.

REFERENCES

1. S lordanescu. *Mol Gen Genet 241*:185–192, 1993.
2. LE Alksne, SJ Projan. *Curr Opin Biotechnol 11*:625–636, 2000.
3. JF Barrett, JA Hoch. *Antimicrob Agents Chemother 42*:1529–1536, 1998.
4. JJ Hilliard, R Goldschmidt, L Licata, EZ Baum, K Bush. *Antimicrob Agents Chemother 43*:1693–1699, 1999.
5. CA Cummings, DA Relman. *Emerg Infect Dis 6*:513–525, 2000.
6. MB Eisen, PT Spellman, PO Brown, D Botstein. *Proc Natl Acad Sci USA 95*: 14863–14868, 1998.
7. MB Eisen, PO Brown. *Methods Enzymol 303*:179–205, 1999.
8. P Fawcett, P Eichenberger, R Losick, P Youngman. *Proc Natl Acad Sci USA 97*: 8063–8068, 2000.
9. DP Zimmer *et al. Proc Natl Acad Sci USA 97*:14674–14679, 2000.
10. S Peterson, RT Cline, H Tettelin, V Sharov, DA Morrison. *J Bacteriol 182*: 6192–6202, 2000.
11. R Rappuoli. *Proc Natl Acad Sci USA 97*:13467–13469, 2000.
12. RJ Lipshutz, SP Fodor, TR Gingeras, DJ Lockhart. *Nat Genet 21*:20–24, 1999.
13. PM Dunman *et al.* In Press *J Bacteriol* 2001.

14. PM Dunman *et al.* (Manuscript in preparation).
15. A de Saizieu *et al. J Bacteriol 182*:4696–4703, 2000.
16. A Dove. *Nat Biotechnol 17*:233–236, 1999.
17. AQ Emili, G Cagney. *Nat Biotechnol 18*:393–397, 2000.
18. RD Smith. *Nat Biotechnol 18*:1041–1042, 2000.
19. VK Singh, RK Jayaswal, BJ Wilkinson. *FEMS Microbiol Lett 199*:79–84, 2001.
20. SP Gygi, GL Corthals, Y Zhang, Y Rochon, R Aebersold. *Proc Natl Acad Sci USA 97*:9390–9395, 2000.
21. C Gauss, M Kalkum, M Lowe, H Lehrach, J Klose. *Electrophoresis 20*:575–600, 1999.
22. AJ Butte, P Tamayo, D Slonim, TR Golub, IS Kohane. *Proc Natl Acad Sci USA 97*:12182–12186, 2000.
23. DA Israel *et al. J Clin Invest 107*:611–620, 2001.
24. M Kuroda *et al. Lancet 357*:1225–1240 2001.

13

Using Fungal Genomes for the Discovery, Development, and Clinical Application of Novel and Current Antifungal Therapeutics

Todd A. Black, Joan K. Brieland, Jonathan Greene, Catherine Hardalo, and Scott S. Walker

Schering-Plough Research Institute, Kenilworth, New Jersey, U.S.A.

Guillaume Cottarel

Genome Therapeutics Corp., Waltham, Massachusetts, U.S.A.

INTRODUCTION

Invasive fungal infections are important causes of morbidity and mortality, especially among immunocompromised patients. These infections, caused by fungal pathogens such as *Aspergillus* species, *Blastomyces dermatitidis*, *Candida* species, *Coccidioides immitis*, *Cryptococcus neoformans*, *Histoplasma capsulatum*, *Fusarium* spp., *Sporothrix schenckii*, the Zygomycetes, or the dematiaceous fungi (agents of phaeohyphomycosis) are among the most difficult infectious diseases to treat. By 2009, the number of serious fungal infections is predicted to grow to over 540,000 cases, up from an estimated 455,000 cases in 1999 (1) (Figure 1). In addition, fluconazole resistance in *Candida* species, as well as incidence of infections due to filamentous fungi resistant to polyenes (e.g., *Fusarium*), are increasing, especially in immunosuppressed patients who may require long-term courses of prophylaxis or chronic maintenance therapy (2,3). In 1996 in the United States alone, there were 10,190 aspergillosis-related hospitalizations, resulting in

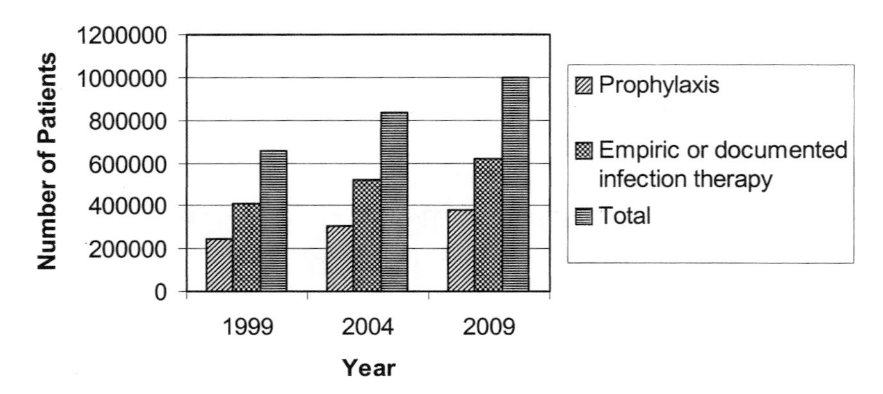

FIGURE 1 The actual patient population in the United States receiving antifungal therapy in 1999 and the predicted increase in the U.S. patient population receiving antifungal therapy for years 2004 and 2009 are shown. The numbers were derived from a survey and report entitled *Strategic Overview of Fungal Infections* prepared by Decision Resources, Inc. of Waltham, MA (see 1).

176,272 hospital days, $633 million in costs, and 1970 deaths (4). The azole antifungal agents (e.g., Fluconazole) have been the mainstay of the current antifungal armamentarium due to their favorable toxicity profile (compared to amphotericin B) and the availability of oral and parenteral formulations. Despite the development of less toxic antifungal agents such as azoles and lipid formulations of amphotericin B, success with long courses of treatment may be limited due to toxicity or the emergence of resistance. As recent data from studies of aspergillosis show, overall treatment failure rates are 36% and only 22% of severely immunosuppressed patients have a complete response to treatment (5). The high failure rates for antifungal therapy are due to several factors. Perhaps the primary factor is the patient population requiring therapy. Extreme or chronic immunosuppression creates a predisposition for serious invasive infections and exposes a need for fungicidal activity to clear these infections. Amphotericin B, considered one of the best fungicidal agents, suffers from a narrow therapeutic index where efficacy may be hindered by toxic side effects. As a result, there is clearly an unmet medical need for more potent, broader spectrum antifungals.

The targets of action of current antifungal therapeutics are primarily limited to cell membrane and sterol biosynthesis. The emergence of resistance and cross-resistance to these agents is beginning to limit their clinical efficacy. With the introduction of a new class of antifungals, the echinocandins (6), there is the additional target of 1,3-β-D-glucan synthesis, but these agents, like amphotericin B, are parenteral formulations only. Thus, there is a continued need to identify new antifungal agents that can provide novel mechanisms of action; favorable therapeutic indices; and potent, fungicidal activity for treatment of serious fungal infections.

The utilization of genomics-based approaches has advanced the search for novel and specific antifungal agents by allowing an early evaluation of potential therapeutic targets. Viewing entire genomes of pathogenic fungi allows for target selection based on the spectrum and therapeutic indices for mechanism-based toxicity by examining the conservation of target proteins among pathogens that do not exist in the host. This early phase of target discrimination should alleviate the difficulties associated with most current antifungal therapies that suffer from narrow therapeutic indices. The directed biological evaluation of these targets also allows for selection of functions that are more likely to lead to development of fungicidal compounds.

FUNGAL TARGET DEFINITION THROUGH BIOINFORMATICS

The fundamental concept behind a genomics-based approach to drug discovery is that, given the complete genome sequence of a target organism, all potential target proteins in that organism are accessible. In order to select potential targets from a genome-derived proteome, a clear set of criteria are required that will define attractive targets. In the case of targets for antifungal drug intervention, criteria for ideal targets have been established and are similar to those utilized for antibacterial drug targets (Figure 2). Attractive antifungal protein targets are those proteins whose (1) functions are present as homologs in all fungal pathogens, (2) are not present as homologs in humans, and (3) are critical for the survival of fungal cells. The first step in a genomics-based approach to antifungal drug discovery is the use of bioinformatics to define the list of fungal proteins that satisfy the first two criteria of spectrum and selectivity.

In 1996 the *Saccharomyces cerevisiae* genome became the first fully sequenced fungal genome (7) and remains the only publicly available complete fungal genome. Furthermore, in order to satisfy the final target criterion of proving the target to be required for survival, the amenable genetics of *S. cerevisiae* made it the model of choice to initiate a bioinformatics analysis to derive a preliminary target list. The strategy was relatively straightforward: ORF sequences from *S. cerevisiae* were compared to publicly available higher eukaryotic sequences represented in the mammalian and primate subsections of GenBank, as well as human and mouse Expressed Sequence Tags (EST) sequences from dbEST, human UniGene clusters from NCBI (8,9), and the proprietary human EST collection from

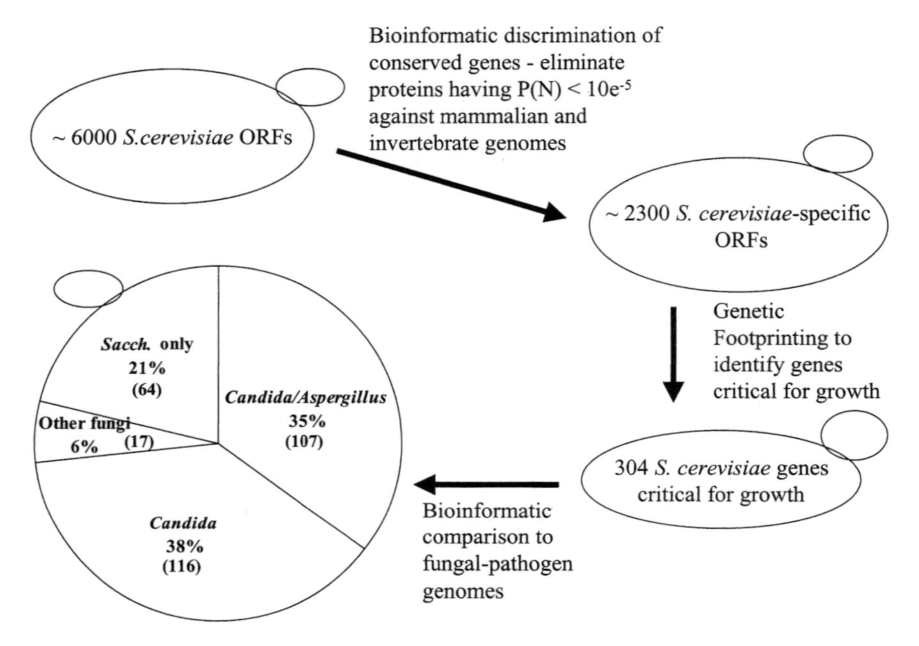

Figure 2 A flow diagram displaying a means to utilize the fully sequenced *S. cerevisiae* genome to derive a focused list of potential antifungal target genes. The *S. cerevisiae* genome was compared with available genomic sequences from higher eukaryotic organisms. Open reading frames (ORFs) that displayed a significant similarity to higher eukaryotic ORFs [$P(N) \leq 10e-5$] were eliminated from the program. Genetic footprinting was then used to identify *S. cerevisiae*-specific ORFs that were critical for growth. The critical *S. cerevisiae*-specific ORFs were subjected to a second round of bioinformatic comparisons with pathogenic fungal sequences to assess the potential target spectrum.

Human Genome Sciences, Inc. *Saccharomyces cerevisiae* ORFs were also compared to available bacterial sequences, including the bacterial subset of GenBank as well as the then-recently completed genomes of *Haemophilus influenzae* and *Helicobacter pylori* (10,11). Comparisons were carried out *en masse* using the BLAST algorithm (12) in conjunction with scripts written in the PERL programming language. Results of BLAST searches were stored using the Sybase SQL Relational Database for subsequent retrieval and analysis using a modified form of a database schema designed by D. Bassett as part of the XREFdb project (13). In 1996, however, there were two major limitations to this strategy: (1) there were no complete higher eukaryotic genomes at that time and therefore potential higher eukaryotic orthologs would be missed and their *S. cerevisiae* cognate would be falsely included in the target list and (2) there was little available sequence from fungal pathogens and so there was no ability to ensure that any *S. cerevisiae* protein on the target list truly had orthologs in fungal pathogens. Nonetheless, genome sequencing efforts of internal and publicly funded research programs were accelerating and expected to fill "gaps" in the comparative databases. Therefore, BLAST searches were continually updated, especially upon completion of higher eukaryotic genomes such as *Caenorhabditis elegans* and *Drosophila melanogaster*, as well as completion of the "rough draft" human genome sequence (14,15). More recently the *Candida albicans* genomic sequence has been derived through public efforts (Stanford Genome Technology Center Web site at http://www-sequence.stanford.edu/group/candida) and a proprietary collection of *Aspergillus fumigatus* EST sequences provided sources for identifying homologous genes in fungal pathogens (see below).

With the BLAST searches being repeatedly updated and with all of the search results cataloged in a relational database, the derivation of the fungal target list had been a dynamic process until quite recently. Results of BLAST searches were typically extracted from the database or updated "on the fly" and deposited into spreadsheets for analysis. *Saccharomyces cerevisiae* genes were sorted based on BLAST homology score cutoffs to exclude sequences showing significant conservation with sequences from higher eukaryotes. These resulting genes could then be examined or sorted further based on homology cutoffs to prokaryotes and eventually fungal pathogens. An important question in this type of analysis is where to set the BLAST score cutoff in determining whether two sequences are in fact homologs. For the initial phases of this analysis, a BLAST $P(N)$ value of $10e-5$ was chosen based on empirical analysis of sample datasets (J. Greene, unpublished information) such that protein pairs were deemed to be homologous if their $P(N)$ score was less than this and were deemed nonhomologous if their $P(N)$ score was greater than this value. One of the benefits of having the results accessible in a dynamically queryable manner is that this value can be altered in either direction as warranted by the particular circumstances. While the initial $P(N)$ value of $10e-5$ may not be perfect, it was an appropriate value from which to commence further analysis. This process was applied with the understanding that the target lists would become more and more refined through the availability of more sequence and biological data. Individual sequence relationships could be explored in detail, where necessary, to determine the apparent presence or absence of homologs in the desirable case of pathogenic fungi or the undesirable case of higher eukaryotes. In practice, this method proved to be useful in determining a preliminary list of potential antifungal drug targets that would be further refined through subsequent bioinformatics analyses and biological studies.

Target Identification

The initial comparative genomic subtractions disclosed that ~45% of the ORFs were specific to *S. cerevisiae*, which represented ~2300 potential target genes. With nearly half the *S. cerevisiae* genome represented following the first-pass selection criterion, it was clear that a rapid and large-scale genetic technique was required to identify those ORFs that affected growth or survival, the final criterion of target selection.

In choosing a methodology it must be acknowledged that the assignment of target ORFs that affect growth or survival can be either broadly or narrowly interpreted. For example, a strict requirement for an ORF function to allow growth in a defined medium, as assessed by a gene disruption, is perhaps the most stringent genetic definition of "essentiality." However, the term "*essentiality*" assumes a different meaning under various growth scenarios. The *S. cerevisiae ade2Δ* locus is classified as nonessential for cells growing on a rich medium and thus may be designated as nonessential for cell proliferation. Whereas disruption of the ADE2 locus by insertion of a transposon element identified ADE2 as essential for cell proliferation on a rich medium when examined in a competitive growth environment (16). Furthermore, as has been shown in the pathogenic fungus *Cryptococcus neoformans*, the ADE2 locus is required for full virulence in a cryptococcal meningitis infection model in rabbits (17). On the other hand, the adhesion factor EPA1 of *Candida glabrata* was shown to be required for adherence to cultured human epithelial cells, yet in a rodent animal model study, there was no difference in initial colonization and persistence between the EPA1 wild-type and null mutant strains (18). Clearly, caution must be used when assessing targets that may affect *in vivo* growth characteristics and virulence.

Open reading frames that are selected as therapeutic targets on an *in vivo* basis should minimally be required for maintenance of an established infection, the scenario most reflective of chemotherapeutic intervention. Unfortunately at this time, there are no high-throughput techniques that allow for this minimal assessment of growth effects on established infections. While signature-tagged mutagenesis (STM) has been applied to many fungal pathogens, including *A. fumigatus* (19), *C. glabrata* (18), and *C. neoformans* (20) allowing the identification of mutant strains that are nonpathogenic, this approach does not discriminate between ORFs that are required for the establishment or maintenance of an infection. Mutants with attenuated virulence, however, can be further analyzed in a directed, low-throughput manner.

Microarray analysis (21), subtractive libraries (22), or SAGE (serial analysis of gene expression) (23) can also be used to identify transcripts associated with virulence or *in vivo* growth (24). These approaches allow for the identification of ORFs that may be exclusively expressed at specific stages of the infection process. These tools are valuable for the elucidation of temporally and spatially regulated developmental events in the pathogen or during the course of antifungal therapy (25,26). However, expression maintained in, or exclusive to, *in vivo* growth during infection does not necessarily indicate that the ORF is an appropriate therapeutic target.

The limitations associated with *in vivo* target validation approaches are readily circumvented by reliance on the most stringent definition of ORF "essentiality." Assuming that any ORF that is clearly essential for growth on a rich medium will also stop the growth of the organism in an established infection, it is possible to readily define the most attractive therapeutic target ORFs. Several high-throughput ORF- disruption and *in vitro* screening methods are available.

The use of mutagenic chemical agents has been the method of choice to identify conditional lethal mutants (27–29). Any random mutagenic method has the disadvantage that mutations are frequently introduced randomly in multiple loci and require further genetic analysis to define the causative mutation. Additionally, it is possible that not all ORFs of interest can be readily mutated to give a conditional lethal phenotype without exhaustive screening. More directed temperature-sensitive mutagenesis approaches have been developed in which a variety of mutagenic PCR procedures are used to amplify and modify a specific ORF (30,31). The extensively mutated ORFs linked to a selectable marker are then recombined into the chromosome (32). Transformed cells can then be rapidly screened for a temperature-sensitive growth phenotype. Complementation of temperature-sensitive alleles with the ORF of interest verifies that the conditional lethal allele is recessive and corresponds to the appropriate gene. While the high efficiency of genetic recombination in *S. cerevisiae* makes a broad conditional lethal screen feasible, this approach still requires extensive screening and follow-up studies to verify the identity of the targets.

Directed chromosomal deletions have been extensively used to identify ORFs required for growth. One-step gene deletions in *S. cerevisiae* can be readily generated using oligonucleotide primers that have homology (as little as 40 bp) to the regions flanking the ORF of interest. These primers are used to amplify a selectable marker and the product is then used to transform a diploid strain (33,34). Longer flanking regions can also be generated using a PCR-based technique, SOE (splicing by overlap extension), or recombinant PCR (35–37). With either method, the short flanking sequences provide sufficient regions of homology to direct recombinant insertion into the appropriate chromosomal locus. Haploid cells are then generated by sporulation of the diploid. The inability to propagate haploid segregants carrying the deletion indicates those ORFs that are required for growth. The international deletion project consortium utilized this method to delete practically all of the ORFs in the *S. cerevisiae* genome (38,39). While this approach has been used on a large scale, the generation and subsequent analysis of the diploid strains is time consuming.

An ideal method for the directed, high-throughput examination of growth effects associated with the disruption of specific target was developed in 1995 by Smith et al. (16). Genetic footprinting by transposon mutagenesis (16,40) provides a means to disrupt virtually every ORF within a population of cells by generating, on average, one transposon insertion per cell within the initial population (T_0). Repression of transposition and subsequent outgrowth of the T_0 population allows for a comparison of those transposition events (i.e., disrupted ORFs) that affect growth rates. Population analysis involves extraction of chromosomal DNA at various time points and separate PCR amplification reactions using a labeled transposon primer and an ORF-specific primer. Polymerase chain reaction products are then analyzed on standard DNA sequencing gels. At T_0, all ORFs should display a variety of PCR products that are indicative of transposon insertions within multiple locations in the ORF. At later time points, disrupted ORFs that do not affect growth will generate PCR products that are identical to those in the T_0 sample, while ORFs that do effect growth will no longer display PCR products that correspond to transposon insertions within the ORF, generating a "genetic footprint." As mentioned earlier in the ADE2 example, genetic footprinting is inclusive in that disruption of those ORFs that cause a decrease in growth rate, and not necessarily eliminate growth, will be identified with this approach. Nonetheless, given the speed at which many genes can be examined, genetic footprinting is an excellent first-pass screening tool to reduce the number of potential target ORFs. Our application of genetic footprinting to the ~2300 ORFs defined by bioinformatics identified ~350 disrupted ORFs that significantly affected growth. Directed disruption of these 350 ORFs using the PCR-based methods mentioned above showed that ~300 of these ORFs were absolutely required for growth under our assay conditions. Consequently, the target ORFs that are most likely to lead to screening and development of new antifungal agents were efficiently distilled from a larger list of potential targets.

The reduced list of potential targets satisfy the criteria of specificity and requirement for growth; however, because no other complete fungal genomic sequences were available, the spectrum of these potential target ORFs could not be assessed. *Candida albicans* was in the process of being sequenced both privately and publicly and some comparisons could be made against the available sequence for this pathogenic yeast. In order to look beyond the yeasts and to provide an additional database for comparison, *A. fumigatus* was selected as the organism of choice for an EST sequencing effort. An EST library approach focuses on the expressed ORFs in the genome without iterative sequencing of nonexpressed and repetitive elements. Nevertheless, the relatively short sequence reads associated with ESTs allow for the identification of orthologous ORFs in an Euascomycetes. Total RNA was extracted from a clinical isolate of *A. fumigatus*, strain ATCC 201795 (41a), at various stages of germination and hyphal growth to obtain a broad representation of expressed genes. Following production of the cDNA, a direct genomic hybridization was performed to normalize the EST library and reduce the redundancy in the EST sequence collection (41b). The sequences derived from the normalized library were continuously compared with the target ORFs identified from *S. cerevisiae* as well as the Genbank nonredundant database. Sequencing was continued for as long as a linear increase was observed in the number of sequence reads versus the target or GenBank databases. When this relationship began to flatten, it was assumed that most of the ORFs represented in the EST library were covered by sequence reads (Figure 3).

The more recent release of high-fold coverage of the *C. albicans* genome allows the analysis of overall conservation of fungal genes that are required for

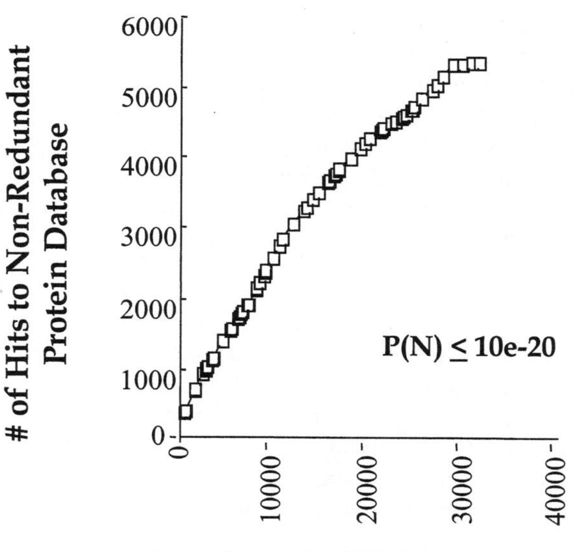

FIGURE 3 A "real-time" value assessment for continued sequencing of a normalized EST library for *A. fumigatus*. cDNA library pools were generated for various growth stages of *A. fumigatus*. The cDNA libraries were then combined and a genomic DNA-driven normalization was performed to provide equivalent representation of all expressed genes. High-throughput sequencing was then performed with the normalized cDNA library. Sequence reads were translated and compared with the nonredundant peptide database at NCBI using a BLAST cutoff criterion of $P(N) \leq 10e-20$. The linear increase in database representation indicates that the cDNA library continues to provide novel sequences without excess sequence redundancy. When the comparison curve began to plateau, further sequencing was ceased as the cDNA library was assumed to be fully mined for novel sequences.

the growth of *S. cerevisiae*. As seen in Figure 2, 107 ORFs were identified as being conserved in both *C. albicans* and *A. fumigatus*, while 116 ORFs were yeast specific and 17 ORFs had hits against other fungal genomes in the public databases, but were not found in *C. albicans* or *A. fumigatus*. In the end, only 64 of the ORFs identified that were required for growth were present only in *S. cerevisiae* (Figure 2). Therefore, the target identification approach described here allowed not only the identification of possible targets for antifungal drug discovery, but also an ability to prioritize the targets based on the effective spectrum of the target.

TARGETED ASSAY DEVELOPMENT

Even with the stringent distillation of potential target genes, the hundreds of possible targets present a formidable challenge for further prioritization and screening for inhibitors. At the forefront of this challenge is the fact that many of the targets have no known or readily assayable biochemical function. The rapid advancement of these targets into drug discovery relies on the broad exploitation of genetic, molecular, biochemical, and physiological characteristics of the fungal cell or through advances in a variety of generic physical-chemical methods for cell-free screening of proteins without a described function.

Cell-Based Assays

Antifungal drug discovery has been based entirely on identifying and characterizing novel chemistries that are active against whole cells. This approach has been fruitful for discovery and evaluation of a wide array of mechanistically distinct classes of potent antifungal agents (42,43). Whole-cell assays also have the obvious advantage over cell-free assays in identifying chemistries that have already achieved penetration into these complex-walled organisms. Taking advantage of whole-cell screens in a targeted manner may allow for a focused effort on chemistries that have a greater potential to become drugs. Modifying a nonspecific cell-based assay into a target-specific high throughput screen (HTS) has been achieved by several means. These cell-based or "genetic screens" rely on the change in the potency of specific compounds in relation to the level of expression of specific target genes.

Compound screening against an array of conditional (e.g., temperature-sensitive) mutant strains offers an example of producing target-specific assays (44). Screening is performed at a semipermissive temperature to achieve a state where the amount or function of a mutated target protein makes the cell particularly vulnerable to growth inhibition. The comparison with conditional mutations in other targets and/or a wild-type strain show that the compound/target relationship is specific. Production of an array of conditional lethal alleles in *S. cerevisiae* can be performed in a directed manner, as described previously for target identification.

Manipulation of the dosage of a target gene can also be used to generate an array of screening strains. *Saccharomyces cerevisiae* stains that are heterozygous for the gene of interest have half the gene dosage of that target when compared to a wild-type or a heterozygous mutant of another target. This reduced gene dosage manifested, as underexpression of a target relative to the wild type, may render the cell more sensitive to the effect of an inhibitor of that gene product. Giaever et al. (45) demonstrated that induced haploinsufficiency could be utilized to identify the targets of known antifungal agents. This concept could be turned into a drug discovery screen by assembling a panel of heterozygous mutants for the targets of interest. A target/compound relationship is again identified as a specific heterozygote that shows hypersensitivity to a compound or extract. Assembling a collection of target heterozygotes for *S. cerevisiae* has been facilitated by the commercial availability of mutants from the *Saccharomyces* Genome Deletion Project (www.resgen.com and 38). Adaptation of this approach to an asexual diploid fungal strain, such as *C. albicans*, is readily feasible using gene-disruption systems that are currently available (see 46 and references therein). Inducible antisense RNA constructs may also be used in a similar manner (47).

While it is the converse of an underexpression assay, overexpression of a target gene may also serve as the basis to identify specific target/compound interactions for libraries of compounds displaying antifungal activity. For example, overexpression of the yeast ERG11, the gene encoding lanosterol carbon-14 demethylase, conferred resistance to the lanosterol carbon-14 demethylase inhibitors, ketoconazole, and miconazole (48). The increased production of the target protein in the cell appears to titrate the inhibitory activity of the compound. The commercial availability of yeast genes in a high-copy-number expression plasmid (www.resgen.com) and the growing knowledge of expression systems in other fungi makes overexpression rescue screening quite tractable.

Cell-Free Assays

The broadest access to potentially useful therapeutic compounds is acquired through cell-free screening. Eliminating the requirement for activity against the cell increases the potential to identify novel compounds that specifically interact with a target protein. Of course eventually, any successful compound candidates derived from cell-free screening must achieve cell penetration.

Cell-free screens have typically required the development of biochemical assays with endpoint readouts that are compatible with high-throughput detection methods. Therefore, even potential target proteins that have a known biochemical function can be difficult to format for screening. Furthermore, defining the biochemical function of a newly identified target protein may necessitate an immense endeavor, only to find that the function is not amenable for high-throughput screening methods. Fortunately, there have been many advances in biophysical techniques that provide fairly generic screening formats. These techniques allow for the identification of small molecule ligands that interact with the target protein of interest.

Generic biophysical approaches for screening rely on the ability to detect an interaction between a small molecule and a protein of interest. These interactions can be measured as simply as the cosegregation of a protein and a ligand through a size exclusion matrix. Variable lengths and residence time on size exclusion columns can give rough estimations of the affinity of the interaction. The cosegregated complex is then subjected to denaturation, organic extraction, and reverse phase separation of the ligand. For high-throughput drug discovery, mixtures of mass-coded combinatorial libraries can be used so that identification of the ligand can be obtained by mass spectroscopy alone. Screening for ligands by affinity selection from natural product sources is also possible, but the characterization of the active compounds will require more extensive purification and spectroscopic analysis (49).

Capillary electrophoresis is another chromatographic technique that has been utilized for measuring protein/ligand interactions (50). The surface charge-to-mass ratio dictates the mobility of the protein through the capillary and binding of a ligand to a protein will modify this ratio leading to a detectable alteration in mobility. This approach is rapid; requires only small amounts of protein and compound; and appears to be applicable to screening complex mixtures, including crude natural product extracts (50,51).

Monitoring ligand effects on target protein thermal denaturation provides another method to identify ligands of target proteins (52). Recently a series of small molecules has been shown to cause dose-dependent thermal stabilization and functional recovery of a mutant tumor suppressor p53 (53), lending pharmaceutical relevance to this idea. A particularly attractive way of measuring ligand-dependent thermal stabilization exploits changes in fluorescence output of dyes such as NanoOrange or SYPRO orange (Molecular Probes, Inc., Eugene, OR) due to changes in protein structure (54). Differentiation of ligand-bound and unbound targets is quite rapid (seconds), making it appropriate for HTS (54). However, application of this homogeneous, isothermal method is predicated on understanding the denaturation characteristics of each target.

Spectroscopic and optical methods are also applicable to screening for novel chemical ligands. Advances in mass spectroscopy (MS) and nuclear magnetic resonance (NMR) spectroscopy have produced related affinity selection and analysis methods having applications in HTS (55–57). Nuclear magnetic resonance spectroscopy of candidate ligands gives an indication of binding specificity and serves to cluster hits based on chemical shift perturbations (57). Alternatively, surface plasmon resonance (SPR) biosensor technology may facilitate screening of many of the targets generated from a genomewide survey (58). Demonstration of the interactions of HIV protease with small molecules demonstrate the utility of this approach to screening for novel ligands (59).

Other Screening Approaches

Other cell-based and cell-free formats for assay development may also find utility in high-throughput screening. These approaches will require at least some knowledge of target properties and/or involve some target-specific development. For example, whole-cell reporter assays (60) could be built for specific targets through the use of transcriptional profiling (i.e., microarrays) or proteome analysis and with some hint about target function a biosynthetic pathway screen may be feasible. Alternatively, yeast two-hybrid (61) or phage display methodologies (62) may identify interacting peptides to allow a function-blind, high-throughput displacement assay.

Genomics for Fungal Diagnostics and Analysis of Host Responses

Conventional diagnostic methodologies for detection of fungal pathogens include *in vitro* cultivation, serological analysis of host-derived antibodies or fungal-derived antigens, and/or direct microscopic detection of fungal organisms in tissues or fluids. These methodologies have significant limitations with regard to sensitivity and specificity, with successful detection frequently limited to patients with advanced stages of disease (63). In contrast, application of sequence-based microbial molecular diagnostic methodologies, including PCR technology, facilitates direct detection of fungal DNA in blood or otherwise normally sterile body fluids and requires only a relatively short processing time and yields a high level of sensitivity and specificity (64–72). Because of the ability to amplify extremely small quantities of fungal DNA, PCR offers earlier detection of fungal pathogens, facilitating rapid initiation of antifungal therapy, thereby potentially improving patient survival (64)].

Polymerase-Chain-Reaction-based diagnostics for detection of medically important fungi most commonly employ universal primers that are complimentary to conserved regions of a particular gene shared by all fungi (63,73). Subsequently, the specific fungal genus and species can be determined by several different methodologies, including direct sequencing of the amplified product, hybridization of the amplicon with a specific probe [e.g., Southern hybridization, slot blotting, probe ELISA, hybridization protection assays, TaqMan analysis (69,74)], restriction fragment length polymorphism (RFLP) analysis (often based on variations in ribosomal genes or the mitochondrial genome) (75,76), high-density chip microarrays (77), and/or single-stranded conformational polymorphisms [SSCP (64)]. Each of the above methods has liabilities and limitations.

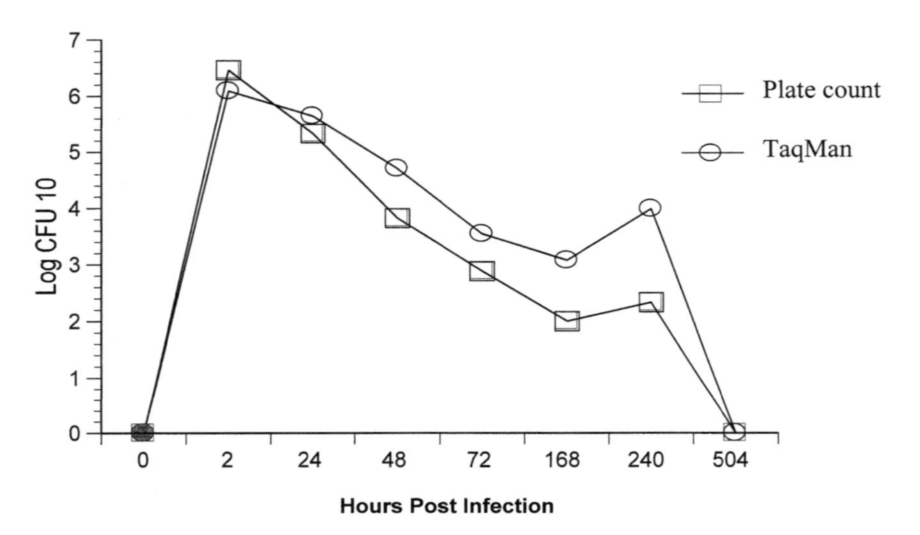

FIGURE 4 A comparative assessment of *A. fumigatus* burden in a mouse pulmonary infection model. Immunocompetent mice were infected with $\sim 10^6$ spores by inhalation on day 1. The fungal burden was assessed by plate counts and TaqMan analysis of bronchial alveolar lavage fluid. For the TaqMan analysis, CFU values were obtained from a standard curve derived by comparing the threshold cycle time required to visualize the specific amplification of the *alp* gene for know quantities of *A. fumigatus* CFUs.

While sequencing of PCR products is the most rigorous method of identification, it is also the most time consuming. Hybridization analysis requires a series of probes, while the use of restriction enzymes may fail to distinguish between subtle variations in sequences. Likewise the pattern of fragment bands observed with SSCP is dependent on conditions employed during sample electrophoresis.

In addition to diagnosis, recent studies have demonstrated a promising role for PCR-based methodology for monitoring microbiological outcome in response to therapy (Figure 4). While PCR-based methodology to evaluate antifungal efficacy has been largely unexplored, it has been used successfully in evaluation of antibiotic efficacy against bacterial pathogens, including *Mycobacterium tuberculosis* and *Treponema pallidum* (78,79). In these studies, patient samples (sputum and cerebrospinal fluid respectively) were taken prior to and at various intervals after antibiotic treatment. The samples were analyzed for the presence of *T. pallidum* or *M. tuberculosis* using PCR with nested primer pairs based on the DNA sequence of the 39-kDa *bmp* gene of *T. pallidum* and IS*6110* insertion sequence specific for the *M. tuberculosis* complex respectively (78,79). Results of these studies demonstrated (1) a high correlation between results of molecular diagnostics and more conventional methodology (i.e., culture results), (2) the correlation of successful therapy and a negative PCR, (3) earlier availability of PCR results when compared to conventional methodology, and (4) the utility of PCR methodology for detecting relapses. This rapid evaluation of treatment efficacy provided by PCR-based diagnostic methodology facilitates rational choices with regard to continuation and/or changes in antibiotic therapy in difficult cases as well as obviating gaps in therapy, thereby potentially improving patient outcome.

Polymerase-Chain-Reaction-based diagnostic techniques are also useful in evaluating population transitions and emergence of resistance in response to antifungal therapy. The wide use of antibiotics, the development of organ and bone marrow transplantation, and the increasing number of immunocompromised patients have resulted in a dramatic increase in both the prophylactic and therapeutic uses of antifungal agents, particularly azoles (80). The extensive use of azole therapy has resulted in a shift in etiology of invasive fungal diseases, including candidemia. There has been a decline in infections due to azole-susceptible *C. albicans* and a steady increase in disease caused by azole-resistant non-*C. albicans* spp., including *C. glabrata* (81–83). In a given patient, the population transition from an azole-susceptible to an azole-resistant *Candida* strain can result from *in vivo* selection of a resistant isolate within a double population of yeasts (one susceptible and one resistant) or, alternatively, acquisition of resistance in a clonal isolate (84–86). Acquired clonal resistance in *Candida* isolates can be mediated by point mutations in the drug target leading to reduced affinity of the target enzyme for the drug (87–91), overexpression or amplification of the target gene (85,86,92,93), mutations in genes encoding other enzymes involved in the biosynthetic pathway (94), and/or enhanced efflux of the drug due to increased expression/activity of membrane proteins responsible for multidrug resistance [ATP or major facilitator transporter superfamilies (92,93,95–99)]. The use of methodology, including pulse-field gel electrophoresis, restriction fragment length polymorphism analysis, Southern blotting with repeat sequence probes, random amplified polymorphic DNA, and interrepeat PCR can delineate clonal relationship between pre- and posttreatment fungal isolates as well as facilitate identification of the molecular mechanism(s) of acquired resistance (100).

These molecular techniques are also useful in evaluating host-pathogen interactions *in vivo*. There has been recent identification of cytokine networking *in vivo* in biologically relevant target organs in animal models of invasive fungal diseases, including pulmonary aspergillosis and systemic candidiasis (41,101).

The effect of fungal infection on induction of cytokine specific mRNAs and the corresponding immunoreactive proteins in relevant target organs can be determined using reverse transcriptase–PCR and cytokine specific ELISAs respectively. Subsequently, using appropriate cytokine specific monoclonal antibodies and/or cytokine specific knockout mice, the biological relevance of cytokines induced in response to a fungal pathogen can be assessed. It is anticipated that results of these studies will provide a rational approach to the development of adjuvant therapy, including immunotherapeutics for treatment of fungal diseases that respond poorly to classical antifungal therapy.

While use of PCR-based methodology in infectious diseases offers enhanced sensitivity and specificity over conventional diagnostic methods; it is not without limitations. Potential complications include (1) false-positive reactions due to contaminating DNA; (2) the inability to distinguish between live and dead organisms in patient samples (102); and (3) false-negative reactions due to human error, low target or poor amplification, and/or the presence of interfering substances in biological samples (103). However, with regard to differentiating sterilization from development of a chronic infection with a positive PCR results, recent studies suggest that PCR positivity to dead organisms is much shorter lived than is a positive response to live infection. A serial study of PCR-positive tissues and fluids and/or multiple samples for PCR analysis may be particularly useful in enhancing sensitivity and providing conclusive results (79). Despite these potential pitfalls, PCR-based diagnostics for fungal diseases, once considered only a research tool, are finding a niche in many clinical microbiology laboratories, offering the hope of rapid diagnosis and directed therapy for fungal infections.

SUMMARY

The growing availability of genomic sequences, powerful bioinformatics, and genetic tools has heralded a new era in antifungal drug discovery and fungal diagnostics. With potentially hundreds of targets generated from a genome-wide survey, effectively directing drug discovery at targets and not the whole-cell relies on our ability to screen many targets in a variety of formats and against a diverse array of chemical sources. In addition, a vigorous clinical assault on a fungal infection requires an accurate diagnosis; here too, genomics will play a role. These developments will no doubt prove to be essential in the ongoing struggle to diagnose and combat the rising occurrence of these debilitating and increasingly recalcitrant infections.

ACKNOWLEDGEMENTS:

Gracious thanks to Drs. Karen Shaw, George Miller, Roberta Hare, George "Skip" Shimer, Jerry Vovis, and Don Moir, who served as directors and mentors to many of the authors. Also thanks to Drs. Beth DiDomenico, Marco Kessler, Debbie Willins, and Jörk Nölling for their scientific input. Drs. Qiandong Zeng and Luquan Wang provided bioinformatics support to our projects. Also thanks to Craig Jackson for allowing us to publish his preliminary data on TaqMan-based diagnostics for *A. fumigatus*.

REFERENCES

1. Decision Resources Inc. *Strategic Overview of Fungal Infections*, 2000.
2. KA Marr. *Curr Opin Infect Dis 13*:615–620, 2000.
3. Working Party for the British Society for Antimicrobial Chemotherapy. *J Antimicrob Chemother 32*:5–21, 1993.
4. EJ Dasbach *et al. Clin Infect Dis 31*:1524–1528, 2000.
5. TF Patterson *et al. Medicine 79*:250–260, 2000.
6. GM Keating, B Jarvis. *Drugs 61*:1121–1129; discussion 1130–1131, 2001.
7. A Goffeau, BG Barrell, H Bussey, RW Davis, B Dujon, H Feldmann, F Galibert, JD Hoheisel, C Jacq, M Johnston, EJ Louis, HW Mewes, Y Murakami, P Philippsen, H Tettelin, SG Oliver. *Science 274*:563–567, 1996.
8. MS Boguski, TM Lowe, CM Tolstoshev. *Nat Genet 4*:332–333, 1993.
9. GD Schuler. *J Mol Med 75*:694–698, 1997.
10. JF Tomb *et al. Nature 388*:539–547, 1997.
11. RD Fleischmann, MD Adams, O White, RA Clayton, EF Kirkness, AR Kerlavage, CJ Bult, JF Tomb, BA Dougherty, JM Merrick *et al. Science 269*:496–512, 1995.
12. SF Altschul, TL Madden, AA Schaffer, J Zhang, Z Zhang, W Miller, DJ Lipman. *Nucleic Acids Res 25*:3389–3402, 1997.
13. DE Bassett Jr, MS Boguski, F Spencer, R Reeves, M Goebl, P Hieter. *Trends Genet 11*:372–373, 1995.
14. The *C. elegans* Sequencing Consortium. *Science 282*:2012–2018, 1998.
15. MD Adams, SE Celniker, RA Holt, CA Evans, JD Gocayne, PG Amanatides, SE Scherer, PW Li, RA Hoskins, RF Galle. *Science 287*:2185–2195, 2000.
16. V Smith, D Botstein, PO Brown. *Proc Natl Acad Sci USA 92*:6479–6483, 1995.
17. JR Perfect, DL Toffaletti, TH Rude. *Infect Immun 61*:4446–4451, 1993.
18. BP Cormack, N Ghori, S Falkow. *Science 285*:578–582, 1999.
19. JS Brown, A Aufauvre-Brown J Brown, JM Jennings, H Arst, DW Holden. *Mol Microbiol 36*:1371–1380, 2000.
20. RT Nelson, J Hua, B Pryor, JK Lodge. *Genetics 157*:935–947, 2001.
21. JL DeRisi, VR Iyer, PO Brown. *Science 278*:680–686, 1997.
22. AJ Watson, J Worley, RM Elliott, DJ Jeenes, DB Archer. *Anal Biochem 277*: 162–165, 2000.
23. VE Velculescu, L Zhang, W Zhou, J Vogelstein, MA Basrai, DE Bassett, P Hieter, B Vogelstein, KW Kinzler. *Cell 88*:243–251, 1997.
24. AP Gasch, PT Spellman, CM Kao, O Carmel-Harel, MB Eisen, G Storz, D Botstein, PO Brown. *Mol Biol Cell 11*:4241–4257, 2000.

25. MJ Marton, JL DeRisi, HA Bennett, VR Iyer, MR Meyer, CJ Roberts, R Stoughton, J Burchard, D Slade, H Dai, DE Bassett Jr, LH Hartwell, PO Brown, SH Friend. *Nat Med 4*:1293–1301, 1998.
26. JA Simon, P Szankasi, DK Nguyen, C Ludlow, HM Dunstan, CJ Roberts, EL Jensen, LH Hartwell, SH Friend. *J Cancer Res 60*:328–333, 2000.
27. LH Hartwell. *J Bacteriol 93*:1662–1670, 1967.
28. NR Morris. *Genet Res 26*:237–254, 1976.
29. P Nurse. *Nature 256*:547–551, 1975.
30. RC Cadwell, GF Joyce. *PCR Methods Appl 2*:28–33, 1992.
31. L Minvielle-Sebastia, K Beyer, AM Krecic, RE Hector, MS Swanson, W Keller. *EMBO J 17*:7454–7468, 1998.
32. N Erdeniz, UH Mortensen, R Rothstein. *Genome Res 7*:1174–1183, 1997.
33. A Baudin, O Ozier-Kalogeropoulos, A Denouel, F Lacroute, C Cullin. *Nucleic Acids Res 21*:3329–3330, 1993.
34. A Wach, A Brachat, R Pohlmann, P Phillippsen. *Yeast 10*:1793–1808, 1994.
35. RM Horton, HD Hunt, SN Ho, JK Pullen, LR Pease. *Gene 77*:61–68, 1989
36. R Higuchi, MA Innis, DH Gelfand, JJ Sninsky, TJ White. *PCR Protocols: A Guide to Methods and Applications*. San Diego, CA: Academic Press, 1990. p 177–183.
37. DC Amberg, D Botstein, EM Beasley. *Yeast 11*:1275–1280, 1995.
38. EA Winzeler et al. *Science 285*:901–906, 1999.
39. R Niedenthal, L Riles, U Guldener, S Klein, M Johnston, JH Hegemann. *Yeast 15*: 1775–1796, 1999.
40. V Smith, KN Chou, D Lashkari, D Botstein, PO Brown. *Science 274*:2069–2074, 1996.
41a. JK Brieland, C Jackson, F Menzel, D Loebenberg, A Cacciapuoti, J Halpern, S Hurst, T Muchamuel, R Debets, R Kastelein, T Churakova, J Abrams, R Hare, A O'Garra. *Infect Immun 69*:1554–1560, 2001.
41b. MM Kessler, DA Willins, Q Zeng, RG Del Mastro, R Cook, L Doucette-Stamm, H Lee, A Caron, TK McClanahan, L Wang, J Greene, RS Hare, G Cottarel, GH Shimer Jr. *Fungal Genet Biol 36*:59–70, 2002.
42. B DiDomenico. *Curr Opin Microbiol 2*:509–515, 1999.
43. NH Georgopapdakou. *Curr Opin Microbiol 1*:547–557, 1998.
44. G Miller. Presented at the Fourth International Antifungal Drug Discovery & Development Summit, Princeton, NJ, 1999.
45. G Giaever, D Shoemaker, T Jones, H Liang, E Winzeler, A Astromoff, R Davis. *Nat Genet 21*:278–283, 1999.
46. R. Wilson, D Davis, B Enloe, A Mitchell. *Yeast 16*:65–70, 2002.
47. MD De Backer, B Nelissen, M Logghe, J. Viaene, I Loonen, S Vandoninck, R de Hoogt, S Dewaele, FA Simons, P Verhasselt, G Vanhoof, R Contreras, WH Luyten. *Nat Biotechnol 3*:235, 2001.
48. H Launhardt, A Hinnen, T Munder. *Yeast 14*:935–942, 1998.
49. D Nikolic, S Habibi-Goudarzi, D Corley, S Gafner, J Pezzuto, R van Breemen. *Anal Chem 72*:3853–3859, 2000.
50. YH Chu, C Cheng. *Cell Mol Life Sci 54*:663–683, 1998.
51. I Schneider. *Today's Chem Work* 8, 1999.

52. J Bowie, A. Pakula. Screening method for identifiying ligands for target proteins. United States patent 407945, 1996.
53. B Foster, H Coffey, M Morin, F Rastinejad. *Science 286*:2507–2510, 1999.
54. D Epps, R Sarver, J Rogers, J Herberg, P Tomich. *Anal Biochem 292*:40–50, 2001.
55. S Kaur, L McGuire, D Tang, G Dollinger, V Huebner. *J Prot Chem 16*:505–511, 1997.
56. G Lenz, H Nash, J Satish. Chemical ligands, genomics and drug discovery. *Drug Discov Today 5*:145–156, 2000.
57. F Moy, K Haraki, D Mobilio, G Walker, R Powers, K Tabei, H Tong, M Siegel. *Anal Chem 73*:571–581, 2001.
58. R Rich, D Mysaka. *Curr Opin Biotechnol 11*:54–61, 2000.
59. PO Markgren, M Hamalainen, U Danielson. *Anal Biochem 265*:340–350, 1999.
60. G Dixon, D Scanlon, S Cooper, P Broad. *J Steroid Biochem Mol Biol 62*:165–171, 1997.
61. P Colas, B Cohen, T Jessen, I Grishina, J McCoy, R Brent. *Nature 380*:548–550, 1996.
62. R Hyde-DeRuyscher, L Paige, D Christensen, N Hyde-DeRuyscher, A Lim, Z Fredericks, J Kranz, P Gallant, J Zhang, S Rocklage, D Fowlkes, P Wendler, P Hamilton. *Chem Biol 7*:17–25, 1999.
63. DN Fredericks, DA Relman. *Clin Infect Dis 29*:475–488, 1999.
64. TJ Walsh, A Grancescone, M Kasai, SJ Chanock. *J Clin Microbiol 33*:3216–3220, 1995.
65. TG Buchman, M Rossier, WG Merz, P Charache. *Surgery 108*:338–346, 1990.
66. HA Erlich, D Gelfand, JJ Sninsky. *Science 252*:1643–1651, 1991.
67. SI Fujita, B Lasker, TJ Lott, E Reiss, CJ Morrison. *J Clin Microbiol 33*:962–967, 1995.
68. AR Holmes, RD Cannon, MG Shepherd, HF Jenkinson. *J Clin Microbiol 32*: 228–231, 1994.
69. VL Kan. *J Infect Dis 168*:779–783, 1993.
70. TG Mitchell, RL Sandin, BH Bowman, W Meyer, WG Merz. *J Med Vet Mycol 32*:3351–3366, 1994.
71. C Spreadbury, D Holden, A Aufauvre-Brown, B Bainbridge, J Cohen. *Genomics 5*:874–879, 1993.
72. CM Tang, DW Holden, A Aufauvre-Brown, J Cohen. *Am Rev Respir Dis 148*: 1313–1317, 1993.
73. TJ White, T Burns, S Lee, J Taylor. *PCR Protocols: A Guide to Methods and Applications*. San Diego, CA: Academic Press, 1990, p 315–324.
74. N Kennedy, SH Gillespie, AOS Saruni, G Kisyombe, R McNerney, FI Ngowi, S Wilson. *J Infect Dis 170*:713–716, 1994.
75. RL Hopfer, P Walden, S Setterquist, WE Highsmith. *J Med Vet Mycol 31*:65–75, 1993.
76. M Maiwald, R Kappe, HG Sonntag. *J Med Vet Mycol 32*:115–122, 1994.
77. EA Winzeler, B Lee, JHMcCusker, RW Davis. *Parasitology 118(suppl)*:73–80, 1999.
78. GT Noordhoek, EC Wolters, ME de Jonge, JD van Embden. *J Clin Microbiol 29*: 1976–1984, 1991.

79. JE Claridge, RM Shawar, TM Shinnick, BB Plikaytis. *J Clin Microbiol 31*: 2049–2056, 1993.
80. JP Bouchara, R Zouhair, SL Boudouil, G Renier, R Filmon, D Chabasse, JN Hallet, A Defontaine. *J Med Microbiol 49*:977–984, 2000.
81. PL Fidel, JA Vazquez, JD Sobel. *Clin Microbiol Rev 12*:80–96, 1999.
82. FC Odds. *Int J Antimicrob Agents 6*:141–144, 1995.
83. MH Nguyen, JE Peacock, AJ Morris et al. *Am J Med 100*:617–623, 1996.
84. CA Hitchock, GW Pye, PF Troke, EM Johnson, DW Warnock *Antimicrob Agents Chemother 37*:1962–1965, 1993.
85. H Vanden Bossche, P Marichal, FC Odds, L Le Jeune, MC Coene. *Antimicrob Agents Chemother 36*:2602–2610, 1992.
86. P Marichal, H Vanden Bossche, FC Odds et al. *Antimicrob Agents Chemother 41*: 2229–2237, 1997.
87. H Banden Bossche, P Marichal, J Gorrens, D Bellen, H Moereels, PA Janssen. *Bioch Soc Trans 18*:56–59, 1990.
88. DC Lamb, DE Kelly, WH Schunck et al. *J Biol Chem 272*:5682–5688, 1997.
89. J Loffler, SL Kelly, H Hebart, U Schumacher, C Lass-Florl, H Einsele. *FEMS Microbiol Lett 151*:263–268, 1997.
90. TC White. *Antimicrob Agents Chemother 41*:1488–1494, 1997.
91. D Sanglard, F Ischer, L Koymans, J Billie. *Antimicrob Agents Chemother 42*: 241–253, 1998.
92. D Sanglard, K Kuchler, F Ischer, JL Pagani, M Monod, J Billie. *Antimicrob Agents Chemother 39*:2378–2386, 1995.
93. TC White. *Antimicrob Agents Chemother 41*:1482–1487, 1997.
94. SL Kelly, DC Lamb, DE Kelly et al. *FEBS Lett 400*:80–82, 1997.
95. R Prasad, P De Wergifosse, A Goffeau, E Balzi. *Curr Genet 27*:320–329, 1995.
96. M Goldway, D Teff, R. Schmidt, AB Oppenheim, Y Koltin. *Microbiology 39*: 422–426, 1997.
97. D Sanglard, F Ischer, M Mondol, J Billie. *Microbiology143*:403–416, 1997.
98. ML Hernaez, C Gil, J Pla, C. Nombela. 1998. *Yeast 14*:517–526.
99. H Miyazaki, Y Miyazaki, A Geber et al. *Antimicrob Agents Chemother 42*: 1695–1701, 1998.
100. LF DiFrancesco, F Barchiesi, F Caselli, O Cirioni, G Scalise. *J Med Microbiol 48*: 955–963, 1999.
101. J Brieland, D Essig, C Jackson, D Frank, D Loebenberg, F Menzel, B Arnold, B DiDomenico, R Hare. *Infect Immun 69*:5046–5055, 2001.
102. K Wicher, F Abbruscata, V Wicher, DN Collins, HW Horowitz, I Auger. *Infect Immun 66*:2509–2513, 1998.
103. HJ Burkardt. *Clin Chem Lab Med 38*:87–91, 2000.

14

Genomics in Novel Natural Products Generation

Zhiqiang An and William R. Strohl
Merck Research Laboratories, Merck & Co., Rahway,
New Jersey, U.S.A.

INTRODUCTION

Natural products are important sources of drugs, including antibacterials such as β-lactams, aminoglycosides, tetracyclines, macrolides, and streptogramins (1). Of the 64 antibacterials approved for clinical use between 1983 and 1994, over 78% were unmodified natural products or compounds derived from natural products (2). For more information on natural products and natural product drugs, several comprehensive reviews should be consulted (1–6). Almost all of the natural product drugs were discovered so far through the following steps: collection and identification of natural material, fermentation and extraction, *in vitro* assays for biological activities, isolation and structural elucidation of the active components from the extracts, and finally characterization of the *in vivo* biological efficacy and toxicity profiles. This empirical approach of natural products screening has been and will continue to be effective in the discovery of pharmaceuticals, including anti-infectives. Nevertheless, the recent developments in the genetics of microbial secondary metabolites biosynthesis, genomics, and metabolic engineering are playing an ever-increasing role in facilitating the natural product drug discovery process. A number of extensive, focused reviews on molecular approaches in natural product drug discovery have been published in recent years (5,7–15). This chapter attempts to provide readers with some general concepts and recent progress of applications of genetic engineering and genomics in natural products drug discovery. Due to space limitations, we shall focus on the following two areas: (1) tapping the chemistry potential of unculturable and slow-growing organisms by expressing secondary metabolite-encoding genes or gene clusters in heterologous hosts and (2) generating novel unnatural secondary metabolites by genetically modifying secondary metabolic biosynthesis pathways. The impact of genomics on the biocatalysis and biotransformation of natural products is not discussed in this chapter since a recent special issue of metabolic engineering has been dedicated to the topic (11).

TAPPING THE CHEMISTRY POTENTIAL OF UNCULTURABLE AND SLOW-GROWING ORGANISMS BY EXPRESSING SECONDARY METABOLITE-ENCODING GENES OR GENE CLUSTERS IN HETEROLOGOUS HOSTS

Microbes Are Proven but Essentially Untapped Sources of Natural Products

Efforts to search for bioactive natural products have mainly focused on organisms that can be easily isolated, cultured, and maintained using laboratory media. Approximately 11,900 antibiotic secondary metabolites have been discovered from microorganisms to date and 66% of these were derived from the order actinomycetes, mostly from *Streptomyces* species (4). The number of prokaryotes on Earth is estimated to be $4–6 \times 10^{30}$ cells (16). This large population provides enormous potential for prokaryotes to acquire genetic diversity. Recent estimates for the number of prokaryotic species range from 10^5 to 10^7 (17). Similarly, only 70,000 species of the approximately 1.5×10^6 species of fungi have been described to date (18). It is expected that discoveries of new bioactive secondary metabolites from easily culturable organisms will continue, but expansion of the natural products sources to unculturable and difficult-to-grow microorganisms may lead to new structural classes. Improved cultivation techniques can help to take advantage of some of the less manageable organisms, but the task is likely to become increasingly intractable, and many microorganisms may never be cultivated in the laboratory.

Capturing the Chemical Diversity of Untapped Microbes by Genetic Engineering

One way to capture the genetic diversity among unculturable or slow-growing microorganisms for secondary metabolite discovery is to introduce large segments of genomic DNA from these unculturable microorganisms into genetically amenable host organisms. In this procedure, the host strain is asked to express genes from the donor organism and to produce secondary metabolites that are not being produced by the hosts. These host strains are typically fast growing and easily fermented on a large scale, and genes can be manipulated genetically. For example, *Escherchia coli* has been used as a host to maintain and express environmental DNA isolated from soil (19,20). Figure 1 illustrates this general concept.

Heterologous Gene Expression

The transgenic approach requires that genes or clusters of genes from unculturable organisms be expressed in laboratory strains. This is possible because transcription and translation control sequences from one organism often function in closely related organisms and sometimes even in distantly related organisms. For example, the entire epothilone gene cluster isolated from the Gram-negative myxobacterium *Sorangium cellulosum* was expressed in *Streptomyces coelicolor*, a Gram-positive actinomycete (21). The erythromycin-encoding gene cluster of *Saccharopolyspora erythraea* was engineered into *E. coli* and *S. coelicolor* for heterologous expression and erythromycin production (22,23). It needs to be emphasized that in these cases, promoters from the host strains were used in the heterologous gene expression (21–23) since the gene donor organisms were very different from the heterologous hosts. Nevertheless, it is remarkable that such large and complex gene clusters can be functionally expressed in heterologous hosts. Similarly, the fungal transformation vector pAN7-1 (24), which has the *E. coli* hygromycin phosphotransferase gene (*hph*) under the control of the *Aspergillus nidulans* glyceraldehyde-3-phosphate dehydrogenase (*gpd*) gene promoter and *trpC* terminator, efficiently confers hygromycin B resistance to a variety of fungal species. More than 40 fungal species have been transformed with this vector, ranging from the basidiomycetes *Schizophyllum commune* (25) and *Laccaria laccata* (26) to various ascomycetous species (27). The *A. niger* glucoamylase gene promoter functioning in *Ustilago maydis* (28) is another example of an ascomycete promoter that functions in a basidiomycete. Some *Aspergillus* genes have been isolated by complementation of *Saccharomyces cerevisiae* mutant strains (29). These experiments demonstrated that genes of foreign origin can be expressed in heterologous laboratory strains. However, one must still be cautious when selecting a donor-recipient combination or combinations, as many genes have been shown not to be expressed in heterologous strains using their native promoters and codon biases.

Genes Encoding Natural-Product Biosynthesis Are Usually Clustered in Microbes

Genes involved in the biosynthesis of major microbial secondary metabolites are often clustered. In prokaryotes, essentially all natural products genes discovered to date are found in single clusters. More than 115 bacterial natural products gene clusters have been cloned and analyzed (11). Relatively fewer natural products biosynthesis genes have been cloned from fungi, but it is clear that genes encoding for natural products in fungi also are often clustered. Examples of fungal second-

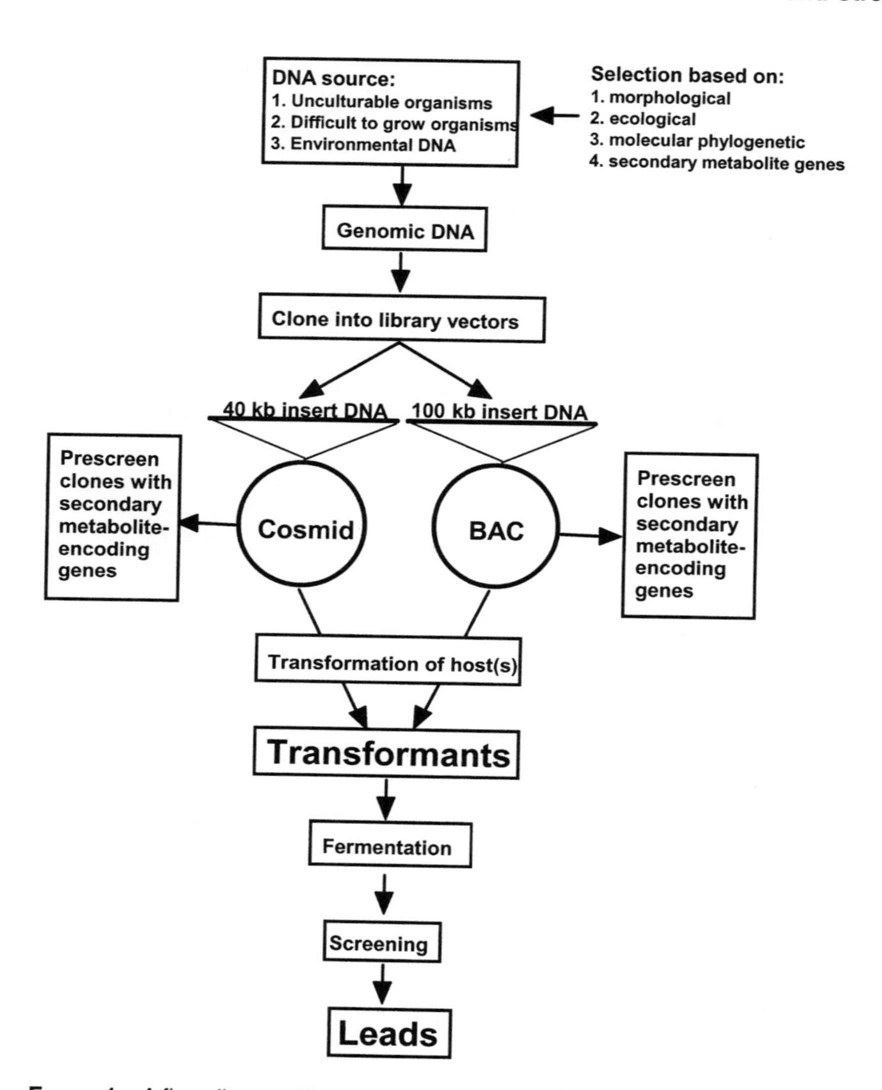

FIGURE 1 A flow diagram illustrates the concept and steps of capturing the chemistry diversity of untapped microbes by heterologous gene expression.

ary metabolite gene clusters are listed in Table 1. This tendency for genes to cluster allows us to gather most, if not all, of the genes for a particular secondary metabolite within one large fragment of DNA which can then be introduced into a recipient laboratory strain. Large pieces of DNA may be cloned from unculturable organisms as cosmid clones (35–45 kb), bacterial artificial chromosome clones (BAC) (>100 kb), or yeast artificial chromosomal clones (YAC) (>100 kb). Alternatively, donor DNA may encode structural enzymes that can be recruited to existing pathways of secondary metabolism in the recipient strain and, as a result, novel secondary metabolites may be produced. Finally, genes encoding transcription and translation regulatory factors that may be introduced to the host genome to either activate silent secondary metabolic pathways in the host or to incite low-activity pathways to biologically and chemically detectable levels. For example, production of the polyketides sterigmatocystin and aflatoxin in *Aspergillus* species is negatively regulated by FadA, the α-subunit of a heterotrimeric G-protein (30–31). In contrast, a dominant activating *fadA* allele (*fadA*^{G42R}) stimulates the transcription of a gene involved in the nonribosomal peptide penicillin biosynthesis in *A. nidulans* (30). Heterologous expression of *fadA*^{G42R} in *Fusarium sporotrichioides* increases trichothecene production (30).

Identifying Productive Transgenic Strains by Prescreening

The transgenic approach to explore genetic diversity among unculturable and slow-growing organisms for secondary metabolites discovery will generate a large number of transgenic strains. If a typical fungal genome is between 30 and 40 Mb (32), about 1000 cosmid clones are needed to have a 1X coverage of a given genome. This translates into about 1000 transgenic strains. It is difficult to estimate the number of secondary metabolic pathways for a given fungus, but a search in Chapman & Hall's *Dictionary of Natural Products* revealed over 35 secondary metabolites produced by *A. nidulans* (33). These metabolites represent at least 12 distinct pathways. Assuming that the average length of a gene cluster for a fungal secondary metabolic pathway is 40 kb and the *A. nidulans* genome is 30 Mb, 12–35 biosynthetic gene clusters constitute about 1.6–4.6% of the genome. Since only a small fraction of a given fungal genome is related to secondary metabolites biosynthesis, the majority of transformants will carry DNA fragments that are not related to secondary metabolite biosynthesis. Bacterial genomes are smaller than fungal genomes, generally ranging from 1 to 5 Mb (34). Some bacterial genomes, however, have much larger genomes. For example, the actinomycete *S. coelicolor* genome is about 8 Mb and at least 12 gene clusters that have been identified or predicted to encode various natural products. These gene clusters constitute about 4% of the entire genome (D. A. Hopwood, personal communication).

TABLE 1 Genes and Gene Clusters Encoding Secondary Metabolites Biosynthesis in Fungi[a]

Fungus	Gene	Product	Activity	Structural type	Reference
A. fumigatus	*alb* 1	Dihydroxynaphthalene	Pigment	Polyketide	(75)
A. nidulans	*pksST*	Sterigmatocystin	Carcinogenic mycotoxin	Polyketide	(76)
A. nidulans	*wA*	Naphthopyrone	Pigment	Polyketide	(77)
A. parasiticus	*pksL* 1	Aflatoxin	Carcinogenic mycotoxin	Polyketide	(78)
A. parasiticus	*pksA*	Aflatoxin	Carcinogenic mycotoxin	Polyketide	(79)
A. terreus	*atX*	6-methyl salicylic acid	Antibiotic	Polyketide	(80)
A. terreus	*LDKS*	Lovastatin (partial)	HMG-CoA reductase inhibitor	Polyketide	(81)
A. terreus	*LNKS*	Lovastatin (partial)	HMG-CoA reductase inhibitor	Polyketide	(82)
Coc. heterostrophus	*pks* 1	T-toxin	T cytoplasm corn toxicity	Polyketide	(83)
Col. lagenarium	*pks* 1	Dihydroxynaphthalene	Melanin	Polyketide	(84)
G. fujikuroi	*fum* 5	Fumonisin	Mammalian toxicity	Polyketide	(85)
G. fujikuroi	*GFPKS4*	Bikaverin	Pigment/antibiotic	Polyketide	AJ278141
Nodulisporium. sp.	*pks* 1	Dihydroxynaphthalene	Melanin	Polyketide	(86)
P. patulum	*msas*	6-methyl salicylic acid	Antibiotic	Polyketide	(87)
Al. alternata	*AMT*	AM-toxin	Phytotoxin	NR peptide	AF184074
A. nidulans	*ACVA*	Penicillin	Antibiotic	NR peptide	P27742
Coc. carbonum	*hts* 1	HC-toxin	Phytotoxin	NR peptide	L48797
Cep. acremonium	*PCBAB*	Cephalosporin	Antibiotic	NR peptide	P25464
G. pulicaris	*pCM* 2	Enniatin	Antibiotic	NR peptide	Z48743
P. chrysogenum	*ACVS*	Penicillin	Antibiotic	NR peptide	P19787, P26046
T. inflatum	*simA*	Cyclosporine	Immunosuppressant	NR peptide	Z28383
U. maydis	*sid* 2	Ferrichrome	Iron chelator	NR peptide	U62738
Cla. fusiformis	*dmaW*	Ergot alkaloid	Neural toxin	Ergot alkaloid	L39640
Cla. purpurea	*d1*	Ergot alkaloid	Neural toxin	Ergot alkaloid	AJ011963
M. roridum	*TRI* 5	Trichothecene	Mycotoxin	Isoprenoid	AF009416
F. sporotrichioides	*TRI* 5	Trichothecene	Mycotoxin	Isoprenoid	M27246
G. fujikuroi	*CPS*	Gibberellin	Phytohormone	Isoprenoid	Y15013
P. roquefortii	*Ari* 1	PR toxin	Mycotoxin	Isoprenoid	L05193

[a] Abbreviations: *A.* = *Aspergillus*; *Al.* = *Alternaria*; *Cla.* = *Claviceps*; *Coc.* = *Cochliobolus*; *Col.* = *Colletotrichum*; *Cep.* = *Cephalosporium*; *F.* = *Fusarium*; *G.* = *Gibberella*; *M.* = *Myrothecium*; *P* = *Penicillium*; *T* = *Tolypocladium*; *U.* = *Ustilago*, *NR* = *Nonribosomal*.

Different approaches can be used to eliminate nonproductive transformants. One is to screen for biological activities. Biological prescreens have to be highly sensitive and high-throughput in order to process a large numbers of extracts. For example, a macrodroplet high-throughput screening technique has been used to identify transgenic strains that produce antibacterial activities (13). The disadvantage of using biological screening is that an assay must be developed for each target. Another approach is to select transformants that contain known secondary metabolite biosynthetic genes before they are sent to screening programs. As discussed earlier, many secondary metabolite-encoding genes have been cloned, sequenced, and characterized (Table 1). As of 1999, 156 microbial secondary metabolite gene clusters had been reported and sequenced and a considerable number has been discovered since then. By using these genes as probes, one can preselect the clones that most likely contain secondary metabolite-encoding gene(s). Similarly, this approach is effective only in a high-throughput mode. Significant progress has been made in the DNA array technology during the past 5 years and it is now practical to apply these technologies to search for secondary metabolite-encoding gene(s) from a large collection of cloned DNA. For example, one method involves creating a secondary metabolite-encoding gene library with consensus DNA fragments obtained from genes encoding various secondary metabolites, such as polyketides and nonribosomal peptides. These DNA fragments can be used individually as probes to screen libraries for secondary metabolite gene clones. They can also be placed on membranes or silicon to create a secondary metabolite-encoding DNA array. One of the limitations to this approach is that the molecular genetics are known for only a few classes of microbial secondary metabolites, mainly polyketides and nonribosomal peptides. Of the 156 microbial secondary metabolites gene clusters reported up to 1999, about 40% were polyketides and 34% were nonribosomal peptides (11). It is well known that in addition to polyketides and nonribosomal peptides, bacteria and fungi also produce shikimates, saccharides, terpenoids, nucleoside, alkaloids, and glycolipids. More secondary metabolite-encoding genes will become available with the continued determination of microbial genomes to come. As of February 2001, there were 40 completed bacterial genomes and an additional 125 bacterial genome projects are in progress (Sanger Centre and TIGR Web sites). Fungal genome projects are relatively few and *Saccharomyces cerevisiae* is still the only complete fungal genome in the public database (35), but there are about 19 eukaryotic microbe genome projects currently under way (Sanger Centre and TIGR Web sites).

Comparison of chemical profiles between the parent strain and the transformants can also be used to separate productive strains from nonproductive strains. Over the past 5 years, there has been an explosion in the use of mass spectral data for high-throughput characterization of organic compounds from small pharmaceuticals to large proteins. The union of a liquid separation technique (LC-MS) with an ionization technique such as electrospray (ESI) has allowed rapid identification of molecular weight and UV characteristics of semipure natural products and other organic molecules (36–38). This and other chemometric tools will continue to be developed to provide higher throughput, increased accuracy, and better reproducibility.

One of the most important tasks in using the transgenic approach to capture natural chemical diversity from the unculturable microorganisms is to rescue and analyze the genes that are responsible for the novel biological activities in a heterologous host. Understanding these new activities at the genetic and molecular levels will enable us to further manipulate the activity. Another major scientific contribution resulting from this approach is the better understanding of heterologous gene expression. Despite recent advances in heterologous gene expression in various microbial systems, our knowledge in the area remains limited. Information on heterologous gene expression will guide scientists to make improved decisions on the selection of unculturable organisms and recipient hosts to maximize the chance of gene expression for the production of novel bioactive compounds.

In summary, by exploring genetic diversity from unculturable and slow-growing microbes and by creating "biocombinatorial" diversity, we improve the likelihood of discovering novel secondary metabolites. Because the recipient strains are fast-growing, "industrial" organisms, once novel drug-producing transformants are identified, scale-up fermentation for commercial production can be implemented using largely traditional approaches.

GENERATING UNNATURAL SECONDARY METABOLITES BY GENETICALLY MODIFYING BIOSYNTHESIS PATHWAYS

Advances in molecular genetics of secondary metabolite biosynthesis have led to a new approach to drug design known as biocombinatorial chemistry. Many academic and industrial laboratories now explore the potential of this genetic engineering/chemistry hybrid approach for developing novel chemicals. Most of the progress so far is in the biocombinatorial synthesis of bacterial polyketides and, to a less extent, nonribosomal peptides.

Polyketides

Polyketides constitute perhaps the largest family of microbial secondary metabolites (39), and the genes for over 60 polyketide biosynthetic pathways from both bacteria and fungi have been cloned and characterized (GenBank database). The controlled variation in chain length, choice of chain-building blocks, and reductive cycle orchestrated by genetically programmed polyketide synthases (PKSs) all lead to the structural diversity of polyketides (40,41). In addition, many PKS products undergo further modification by regiospecific glycosylases, methyltransferases, and oxidative enzymes to produce still greater diversity. The tremendous structural diversity of polyketides reflects the wide variety of their biological properties. Many are effective therapeutic agents, ranging from antibiotics (tetracycline and erythromycin) and anticancer agents (doxorubicin) to immunosuppressants (FK506 and rapamycin). Some polyketides have multiple pharmacological effects; for example, the immunosuppressant FK506 shows antifungal activity (42), and the antifungal compound thysanone also inhibits rhinovirus protease (43). The mode of action of different polyketide drugs varies greatly despite their structural similarities. For example, the immunosuppressants FK506 and rapamycin act very differently on T cells: FK506 inhibits the production of interleukin 2 (44), and rapamycin prevents the proliferative response to interleukin 2 bound at the interleukin 2 receptor (45). Given the diversity of naturally occurring polyketides, plus their clinical and commercial success, it is expected that more effective polyketide drugs will be discovered and developed in many therapeutic areas.

Bacterial Modular PKSs (Type I). Bacterial PKSs are classified into two broad types based on genetic organization and biosynthetic mechanisms (46–48). Modular PKSs (type I) have discrete multifunctional enzymes that control the sequential addition of thioester units, and their subsequent modification, to produce macrocyclic compounds (or complex polyketides). Type I PKSs are exemplified by 6-deoxyerythronolide B synthase (DEBS), which catalyzes the formation of the macrolactone portion of erythromycin A, an antibiotic produced by *Saccharopolyspora erythraea*. There is a total of seven active-site modules in DEBS, with each given module containing three to six active sites. Three domains constitute a minimal module: acyl carrier protein (ACP), acyltransferase (AT), and β-ketoacyl-ACP synthase (KS). Some PKSs contain additional modules for reduction of β-carbons, such as β-ketoacyl-ACP reductase (KR), dehydratase (DH), and/or enoyl reductase (ER). The thioesterase cyclase (TE) protein is present only at the end of module 6 (49).

Bacterial Aromatic PKSs (Type II). Bacterial aromatic PKSs (type II) are composed of several separate, mostly monofunctional proteins, the active sites of which are used iteratively for assembly and functional-group manipulation of the polyketide chain. At least 25 sets of aromatic polyketide PKSs have been cloned from *Streptomyces* and *Saccharopolyspora* species (GenBank database). Best studied among type II PKSs is the PKS for the biosynthesis of the benzoisochromanequinone actinorhodin (40,50–51). Genes for actinorhodin biosynthesis are designated "actI–VII." The actI locus has three genes: KS, AT, and ACP. ActIII encodes KR. ActI and actIII constitute the minimal PKS. The actII locus is responsible for transcriptional regulation of the act genes and for actinorhodin export. The actIV–VII loci encode several postsynthetic modifying functions, e.g., cyclization (VII), aromatization, and subsequent chemical tailoring. Other aromatic PKSs share the same basic architecture with minor structural differences (47,48).

Type II aromatic PKSs contain a single set of iteratively used active sites, although the basic organization of the two types of PKS is similar. These characteristics make it difficult to predict the structure of the polyketide produced after modification of a type II PKS gene. Significant advances in research on the combinatorial biosynthesis of novel aromatic polyketides have been made in laboratories of Hopwood, Khosla, and their collaborators in recent years. Hybrid aromatic polyketides have been generated by transferring partial or complete biosynthetic gene clusters to different polyketide producers, followed by screening for new structures. For example, mederrhodin A, a novel aromatic polyketide, was produced by a medermycin-producing *Streptomyces* strain transformed with the actVA gene (52). This approach of generating novel polyketides was largely empirical, and it required the activity of enzymes for late tailoring steps. During the past few years, tremendous progress has been made in our understanding of genetic programming of aromatic PKSs; consequently, the rational design of novel aromatic polyketides is also advancing. Several novel compounds have been generated by constructing and expressing recombinant PKSs in *Streptomyces* sp. (40,51,53–55). Based on their work and the work of others, McDaniel *et al.* (51) summarized six rules for rational design of novel aromatic polyketides based on chain length, ketoreduction, cyclization of the first ring, first-ring aromatization, second-ring cyclization, and additional cyclization.

Combinatorial Synthesis of Bacterial Polyketides. Pioneered by the Hopwood laboratory (50), molecular genetic analysis of PKS genes has confirmed earlier biochemical and chemical findings that the structural diversity of polyketides is a result of the quantity and types of acyl units involved. This body of research also indicates that novel polyketides can be produced by manipulating the sequence and specificity of enzyme-mediated reduction, dehydration, cyclization, and aromatization through genetic engineering (23,40,41,48,51–59). One of the first examples of the generation of novel polyketides through genetic manipulation of PKS genes was reported in 1990 by the Strohl group (60). They demonstrated the biosynthesis of anthraquinones aloesaponarin II and desoxyerythrolaccin through the cloning of actinorhodin polyketide biosynthesis genes in the aclarubicin-producing strain *S. galilaeus* (60).

The era of rational design for novel antibiotic structures was ushered in by early successes in synthesis of complex polyketides. The modular type I PKSs producing complex polyketides contain a unique active site for each enzyme-catalyzed reaction in the pathway, giving rise to final structures that are determined by the numbers and types of active sites. Donadio *et al.* demonstrated in the early 1990s that analogs of the erythromycin polyketide backbone could be generated by eliminating active sites within the PKS (49,61,62). By repositioning a chain-terminating thioesterase domain from the C-terminus of module 6 of DEBS3 to the C-terminus of module 2 of DEBS1, Cortes *et al.* were able to construct a multienzyme unit that catalyzed only the first two rounds of polyketide chain extension (41). The mutant produced a triketide lactone structure without any trace of the wild-type polyketide erythromycin, indicating premature chain termination and cyclization. By expressing the entire DEBS gene cluster in a heterologous host, substantial quantities of 6-deoxyerythronolide B (the aglycone of the macrolide antibiotic erythromycin) was produced (23).

Bacterial Heterologous Expression Hosts. Combinatorial biosynthesis of novel metabolites requires an efficient expression host and sometimes multiple hosts. Several actinomycetes, mostly *Streptomyces* species, have been used as expression host for combinatorial biosynthesis such as *Streptomyces coelicolor* (51), *Streptomyces lividans* (11), and *Streptomyces venezuelae* (10). Each host–vector system has advantages and disadvantages regarding transformation efficiency, growth rate, and genetic understanding of the host. More recently, erythromycin was produced in *E. coli* (22). This was achieved by multiple genetic engineering steps in the *E. coli* strain, including introduction of the three DEBS genes from *Streptomyces erythraea*, introduction of the *sfp* gene encoding phosphopantetheinyl transferase, deletion of the endogenous *prpRBCD* genes, and overexpression of the endogenous *prpE* and *birA* genes (22). The feasibility of engineering *E. coli* to produce complex natural products provides an alternative host for combinatorial biosynthesis, because *E. coli* can be easily manipulated at the molecular genetic level. The existence of multiple host–vector systems offers the opportunity to select the optimal system for individual projects.

Modifying Enzymes. In addition to manipulating the PKS genes themselves, one can also modify various modifying enzymes such as glycosyltransferaseases, methyltransferases, and oxidative enzymes. For example, manipulation of the desosamine biosynthetic genes of the pikromycin biosynthesis pathway in *S. venezuelae* has led to the isolation of several methymycin/pikromycin analogs (63–65). In a more elaborate experiment, the combination of the pikromycin pathway in *S. venezuelae* and a sugar biosynthetic gene, CalH, from the calicheamicin pathway in *Micromonospora echinospora* yielded two novel macrolides in a *des*I-deleted *S. venezuelae* strain (66).

Fungal Polykeides. Although polyketides are commonly produced by filamentous fungi, nucleotide sequences have been reported for only 14 fungal PKS genes, all of which encode iterative, type I PKSs. Enzymatic domains of a generic fungal PKS-encoding gene are shown in Figure 2. Even though combinatorial manipulation of fungal polyketide biosynthetic genes is not as advanced for purposes of genetic engineering as that of bacterial systems, the interative, yet modular fungal PKS systems exhibit several advantages over bacterial systems (67)

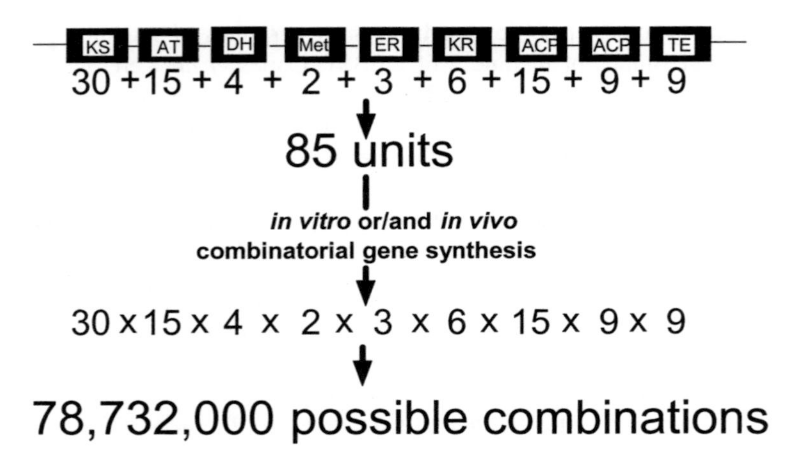

Figure 2 A diagram illustrates the concept of assembling large numbers of chimeric PKS genes by functional domain shuffling. Generic organization of a fungal polyketide synthase gene (PKS) is shown at the top. KS (ketoacyl synthase), AT (acyltransferase), DH (dehydratase), MeT (methyltransferase), ER (enoyl reductase), KR (ketoreductase), ACP (acyl-carrier protein; note that some fungal PKS genes have two different ACP domains), and TE (thioesterase cyclase). Numbers below the functional domain boxes indicate how many representatives of each domain are available from published fungal PKS genes. Assembling the 85 functional domain units in the proper order in all possible combinations would result in 78,732,000 chimeric PKS genes.

as follows: (1) fungal polyketides range in chain lengths from 6 to more than 40 carbons; (2) fungal polyketides can be methylated by a methyltransferase domain that is part of the PKS itself (making it amenable to genetic manipulation), rather than by incorporating propionate as an extender unit as observed in bacteria; (3) fungal polyketides can be partially or completely reduced, and they undergo unusual cyclizations; both features contribute to their potential as starting materials for production of novel polyketides; (4) a considerable amount of biochemical information in a relatively short stretch of DNA because fungal PKSs are both structurally modular and functionally iterative; and (5) fungal PKSs described to date are single proteins (bacterial PKSs often form multiple protein complexes; this characteristic explains why only fungal PKSs are sufficiently stable to be completely purified). In addition, availability of single genes encoding single proteins greatly facilitates heterologous expression of PKSs. Similar principles and roles for genetically engineering bacterial polyketides could be applied to combinatorial synthesis of fungal polyketides, but specific protocols need to be developed for fungi because fungal gene regulation and structure are distinct from those of bacteria. Figure 2 illustrates a concept of assembling large numbers of chimeric PKS genes by functional domain shuffling.

Nonribosomal Peptides and Other Metabolites

A wide variety of peptides is synthesized by nonribosomal peptide synthetases in microbes that exhibit a broad range of biological activities. Such activities may act as antibiotics (penicillins, vancomycin, bacitracin, and gramicidin), siderophores (ferrichrome, mycobactin, and enterobactin), toxins (HC toxin and syringomycin), and immunosuppressive agents (cyclosporin and rapamycin) (9,68). The growing number of nonribosomal peptide synthetase (NRPS) genes isolated and characterized (at least 53 NRPS genes in the GenBank database) is accumulating a wealth of information regarding enzymatic activities and structural features of this family of proteins. One of the most prominent features is modularity (9,69,70). A module, in the context of NRPSs, is defined as a DNA/protein segment that containing all of the functions required for catalysis of peptide elongation and associated substrate modifications such as acylation, epimerization, heterocyclization, and N-methylation of the amide nitrogen. NRPSs can incorporate hundreds of known amino acids into peptide products (71). The order and number of modules on an NRPS determine the sequence and number of amino acids in the final peptide product. The increasing number of NRPS genes cloned and characterized allows us to predict in some cases the substrate specificity of adenylation domains of unknown function (9,68,72). Modular arrangement of NRPSs and substrate specificity of each module in a given NRPS strongly suggests the possibility of creating novel peptide products by genetic and combinatorial manipulations. Although there still are many obstacles in successfully engineering these proteins, recent results have been promising. Working with the three-module surfactin synthetase 1 of *Bacillus subtilis*, Marahiel's group successfully altered the substrate specificity of multimodular peptide synthetases by targeted substitution of minimal-module encoding regions within NRPS genes (73). The same group also constructed functional hybrid peptide synthetases by module and domain fusions (74). In a more recent review, Doekel and Marahiel proposed a strategy to synthesize antipain by genetic engineering (9) and the steps outlined in their strategy summarize the latest development in the combinatorial synthesis of nonribosmal peptides.

SUMMARY

The field of natural products biosynthesis has witnessed exponential progress since the early 1990s. Accumulation of this knowledge has created the new discipline of "combinatorial biosynthesis of unnatural secondary metabolites," but it is only the beginning. It is certain that many more natural products encoding genes and gene clusters will be cloned and characterized in the near future with the completion of more genome sequencing projects. This explosion of genomic information will not only help to provide additional information about the combinatorial synthesis of polyketides and nonribosomal peptides, but it will also accelerate our understanding of many less understood but important metabolic pathways such as terpenoids and alkaloids. The success of tapping the chemical potential of unculturable organisms also relies heavily on the progress of natural products biosynthesis at the biochemical, genetic, and genomic levels.

ACKNOWLEDGMENT

We thank Drs. Jennifer Nielsen Kahn, Ningyan Zhang, and Steven Gould and Ms. Sarah Kennedy for comments on the Chapter.

REFERENCES

1. DJ Newman, GM Gragg, KM Snader. *Nat Prod Rep 17*:215–234, 2000.
2. GM Cragg, DJ Newman, KM Snader. *J Nat Prod 60*:52–60, 1997.
3. AL Demain. *Nature Biotech 16*:3–4, 1998.
4. J Berdy. Proceedings of the Ninth Symposium on the Actinomycetes, Moscow, 1995, pp 13–34.
5. YZ Shu. *J Nat Prod 61*:1053–1071, 1998.
6. WR Strohl, HB Woodruff, RL Monaghan, D Hendlin, S Mochales, AL Demain, J Liesch. *SIM News 15*:5–19, 2001.
7. R Barkovich, JC Liao. *Metab Eng 3*:27–39, 2001.
8. L Du, C Sanchez, B Shen. *Metab Eng 3*:78–95 2001.
9. S Doekel, MA Marahiel. *Metab Eng 3*:64–77, 2001.
10. Y Xue, DH Sherman. *Metab Eng 3*:15–26, 2001.
11. WR Strohl. *Metab Engineer 3*:4–14, 2001.
12. CA Roessner, AI. Scott. *Annu Rev Microbiol 50*:467–490, 1996.
13. WA Wells. *Chem. Biol. 5*:15–16, 1998.
14. KA Thiel. *Bioventure View 13*:1–8, 1998.
15. DE Cane, CT Walsh, C Khosla. *Science 282*:63–68, 1998.
16. WB Whitman, DC Coleman, WJ Wiebe. *Proc Natl Acad Sci USA 95*:6578–6583, 1998.
17. PM Hammond. In: D Allsopp, RR Colwell, DL Hawksworth, eds. *Microbial Diversity and Ecosystem Function*. Wallingford: CAB International, 1995, pp 29–71.
18. DL Hawksworth, AY Rossman. *Phytopathology 87*:888–891, 1997.
19. MR Rondon, SJ Raffel, RM Goodman, J Handelsman. *Proc Natl Acad Sci USA 96*: 6451–6455, 1999.
20. MR Rondon, PR August, AD Bettermann, SF Brady, TH Grossman, MR Liles, KA Loiacono, BA Lynch, IA MacNeil, C Minor, CL Tiong, M Gilman, MS Osburne,

J Clardy, J Handelsman, RM Goodman. *Appl Environ Microbiol 66*:2541–2547, 2000.
21. L Tang, S Shah, L Chung, J Carney, L Katz, C Khosla, B Julien. *Science 287*:640–2, 2000.
22. BA Pfeifer, SJ Admiraal, H Gramajo, DE Cane, C Khosla. *Science 291*:1790–1792, 2001.
23. CM Kao, L Katz, C Khosla. *Science 265*:509–512, 1994.
24. PJ Punt, RP Oliver, MA Dingemanse, PH Pouwels, CAMJJ vandenHondel. *Gene 56*:117–124, 1987.
25. H Mooibroek, AG J Kuipers, JH Sietsma, PJ Punt, JGH Wessels. *Mol Gen Genet 222*:41–48, 1990.
26. V Barrett, RK Dixon, PA Lemke. *Appl Microbiol Biotechnol 33*:313–316, 1990.
27. CAMJJ vandenHondel, PJ Punt, RFM Gorcom. In: JW Bennett, LL Lasure, eds. *More Gene Manipulations in Fungi*. San Diego, CA: Academic Press, 1991, pp 396–428.
28. TL Smith, J Gaskell, RM Berka, M Yang, DJ Henner, D Cullen. *Gene 88*:259–262, 1990.
29. J Rambosek, J Leach. *CRC Crit Rev Biotech. 6*:357–393, 1987.
30. A Tag, J Hicks, G Garifullina, C Ake, TD Phillips, M Beremand, N Keller. *Mol Microbiol 38*:658–65, 2000.
31. K Shimizu, N Keller. *Genetics 157*:591–600, 2001.
32. M Walz. In: U Kuck, ed. *Genetics and Biotechnology*. New York: Springer–Verlag, 1995, vol. II, pp 61–73.
33. Chapman and Hall. *Dictionary of Natural Products on CD-ROM*. Washington, DC: CRC Press, 2000.
34. CM Fraser, JA Eisen, SL Salzberg. *Nature 406*:799–803, 2000.
35. JW Bennett. *Fungal Genet Biol 21*:3–7, 1997.
36. MS Lee, EH Kerns. *Mass Spectrom Rev 18*:187–279, 1999.
37. KG Owens. *Appl Spectrosc Rev 27*:1–49, 1992.
38. O Fiehn, J Kopka, P Dormann, T Altmann, RN Trethewey, L Willmitzer. *Nat Biotechnol 18*:1157–61, 2000.
39. D O'Hagan. *Nat Prod Rep 12*:1–32, 1995.
40. R McDaniel, S Ebert-Khosla, DA Hopwood, C Khosla. *Science 262*:1546–1550, 1993.
41. J Cortes, KEH Wiesmann, GA Roberts, MJB Brown, J Staunton, PF Leadlay. *Science 268*:1487–1489, 1995.
42. T Kino, H Hatanaka, S Miyata, N Inamura, M Nishiyama. *J Antibiot 40*:1256–1265, 1987.
43. SB Singh, MG Cordingly, RG Ball, JL Smith, AW Dombrowski, MG Goetz. *Tetrahedr Lett 32*:5279–5282, 1991.
44. SL Schreiber. *Cell 70*:365–368, 1992.
45. EJ Brown, MW Albers, TB Shin, K Ichikawa, CT Keith, WS Lane, SL Schreiber. *Nature 369*:756–758, 1994.
46. L Katz, S Donadio. *Annu Rev Microbiol 47*:875–912, 1993.
47. DA Hopwood, C Khosla. In: *Ciba Foundation Symposium. Secondary Metabolites: Their Function and Evolution*. Chichester: Wiley, 1992, vol. 17, pp 88–112.

48. CR Hutchinson, I Fujii. *Annu Rev Microbiol 49*:201–238, 1995.
49. S Donadio, L Katz. *Gene 111*:51–60, 1992.
50. F Malpartida, DA Hopwood. *Nature 309*:462–464, 1984.
51. R McDaniel, S Ebert-Khosla, DA Hopwood, C Khosla, *Nature 375*:549–554, 1995.
52. DA Hopwood, F Malpartida, HM Kieser, H Ikeda, J Duncan, I Fujii, BAM Rudd, HG Floss, S Omura. *Nature 314*:642–644, 1985.
53. R McDaniel, S Ebert-Khosla, H Fu, DA Hopwood, C Khosla. *Proc Natl Acad Sci USA 91*:11542–11546, 1994.
54. E Kim, KD Cramer, AL Shreve, DH Sherman. *J Bacteriol 177*:1202–1207, 1995.
55. H Fu, R McDaniel, DA Hopwood, C Khosla. *Biochemistry 33*:9321–9326, 1994.
56. R Pieper, G Luo, DE Cane, C Khosla. *Nature 378*:263–266, 1995.
57. C Khosla, R McDaniel, S Ebert-Khosla, R Torres, DH Sherman, MJ Bibb, DA Hopwood. *J Bacteriol 175*:2197–2204, 1993.
58. T Yu, DA Hopwood. *Microbiology 141*:2779–2791, 1995.
59. C Khosla, S Ebert-Khosla, DA Hopwood. *Mol Microbiol 6*:3237–3249, 1992.
60. PL Bartel, CB Zhu, JS Lampel, DC Dosch, NC Connors, WR Strohl, JM Beale Jr, HG Floss. *J Bacteriol 172*:4816–4826, 1990.
61. S Donadio, MJ Staver, JB McAlpine, SJ Swanson, L Katz. *Science 252*:675–679, 1991.
62. S Donadio, JB McAlpine, PA Sheldon, MA Jackson, L Katz. *Proc Natl Acad Sci USA 90*:7119–7123, 1993.
63. L Zhao, NLS Que, Y Xue, DH Sherman, H Liu. *J Am Chem Soc 120*:12159–12160, 1998.
64. SA Borisova, L Zhao, DH. Sherman, HW Liu. *Org Lett 1*:133–136, 1999.
65. LS Zhao, DH Sherman, HW Liu. *J Am Chem Soc 120*:10256–10257, 1998.
66. L Zhao, J Ahlert, Y Xue, JS Thorson, DH Sherman, H Liu. *J Am Chem Soc 121*: 9881–9882, 1999.
67. DJ Bedford, E Schweizer, DA Hopwood, C Khosla. *J Bacteriol 177*:4544–4548, 1995.
68. GL Challis, J Ravel, CA Townsend. *Chem Biol 7*:211–224, 2000.
69. D Konz, MA Marahiel. *Chem Biol 6*:39–48, 1999.
70. MA Marahiel, T Stachelhaus, HD Mootz. *Chem Rev 97*:2651–2673, 1997.
71. H Kleinkauf, H vonDohren. *Eur J Biochem 192*:1–15, 1990.
72. T Stachelhaus, HD Mootz, MA Marahiel. *Chem Biol 6*:493–505, 1999.
73. A Schneider, T Stachelhaus, MA Marahiel. *Mol Gen Genet 257*:308–318, 1998.
74. HD Mootz, D Schwarzer, MA Marahiel. *Proc Natl Acad Sci USA 97*:5848–5853, 2000.
75. HF Tsai, YC Chang, RG Washburn, MH Wheeler, KJ Kwon-Chung. *J Bacteriol 180*:3031–3038, 1998.
76. JH Yu, TJ Leonard. *J Bacteriol 177*:4792–4800, 1995.
77. ME Mayorga, WE Timberlake. *Mol Gen Genet 235*:205–212, 1992.
78. GH Feng, TJ Leonard. *J Bacteriol 177*:6246–6254, 1995.
79. PK Chang, JW Cary, J Yu, D Bhatnager, TE Cleveland. *Mol Gen Genet 248*:270–277, 1995.
80. I Fujii, Y Ono, H Tada, K Gomi, Y Ebizuka, U Sankawa. *Mol Gen Genet 253*:1–10, 1996.

81. J Kennedy, K Auclair, SG Kendrew, C Park, JC Vederas, CR Hutchinson. *Science 284*:1368–1372, 1999.
82. VA Vinci, MJ Conder, PC McAda, CD Reeves, J Rambosek, CR Davis, LE Hendrickson. U.S. 5744350, 1998.
83. G Yang, MS Rose, BG Turgeon, OC Yoder. *Plant Cell 8*:2139–2150, 1996.
84. Y Takano, Y Kubo, K Shimizu, K Mise, T Okuno, I Furusawa. *Mol Gen Genet 249*: 162–167, 1995.
85. RH Proctor, AE Desjardins, RD Plattner, TM Hohn. *Fungal Genet Biol 27*:100–112, 1999.
86. TR Fulton, N Ibrahim, MC Losada, D Grzegorski, JS Tkacz. *Mol Gen Genet 262*: 714–720, 1999.
87. J Beck, S Ripka, A Siegner, E Schiltz, E Schweizer. *Eur J Biochem 192*:487–498, 1990.

Index